HIDDEN SCENTS

The Language of Smell in the Age of Approximation

Hidden Scents: The Language of Smell in the Age of Approximation
Revised edition, 2016
Allen Barkkume

Cover art by David Günther
Cover design by Theresa Gjertsen
Photograph of young person smelling courtesy of Elsevier Books
Graph of the flavor network courtesy of Barabási Labs
All other artwork by Joe Scordo

ISBN: 978-1-365-29276-7
Smell this book

Contents

Preface

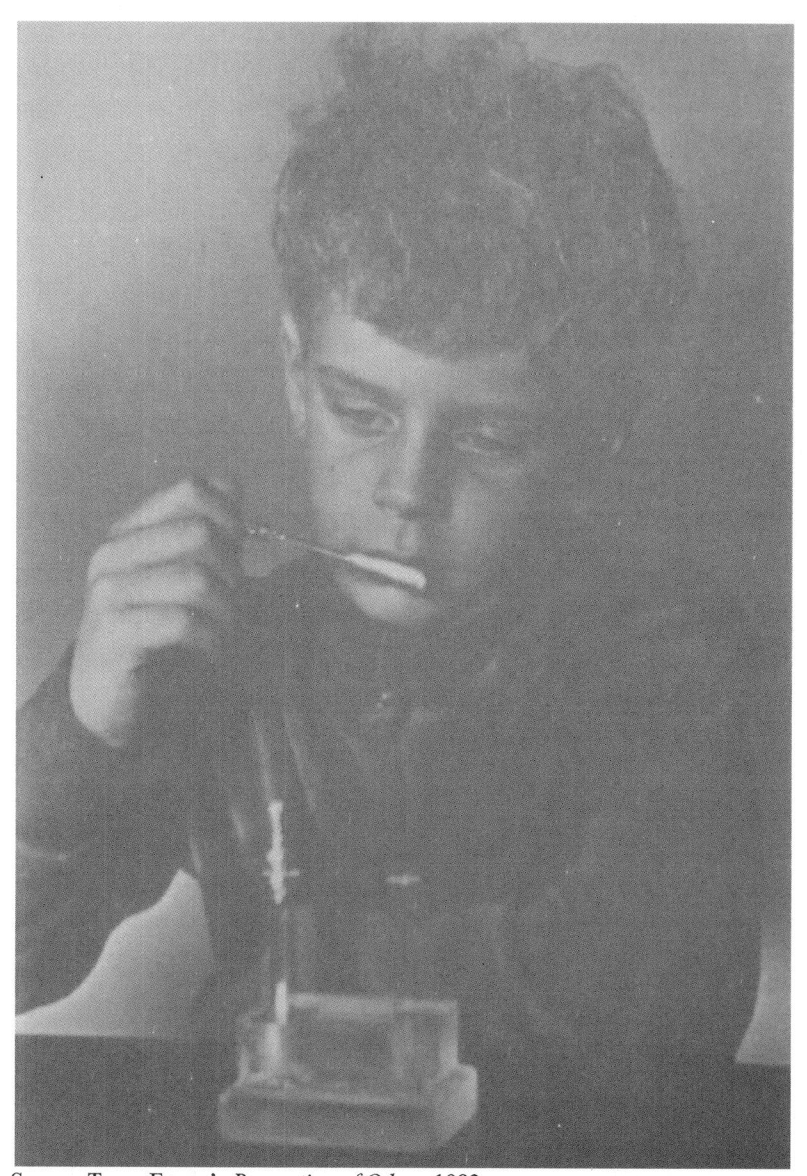

Source: Trygg Engen's *Perception of Odors*, 1982

This boy is smelling. He is not looking at the thing he smells. Rather, he is looking beyond it. With his entire visual system turned off, he has entered the *n*-dimensional world of olfaction. A jungle, he navigates the totality of his experience – emotionally, linguistically, spatially – "I remember this smell, but from where?"

His mind performs a wonder of calculations. His memory, twitching, morphs and re-forms. A wet, oscillating machine simultaneously scans autobiographical spatiotemporal datamaps while performing virtual organic chemistry experiments. It matches every episodic memory of the past eight hours with every potential combination and degradation of chemicals that can be imagined. At the same time, it performs the reverse function: It matches the smell he perceives with every possible episode in which he could have come into contact with it.

In this way he explores the complete interiorization of his external world. His search is not aided by vision, but by something far more effective. Instantiating his virtual body, flashing limbic states, is a haystack of dimensions where every straw is moved ever so slightly in its successive iterations. It is not the physical haystack that he explores, straw-by-straw. Rather, it is the potential configurations of the whole. If an animal could think the way a human thinks, it would be like this.

The following text locates olfactory phenomena at the intersection of information science and the human body. This sense – which is largely hidden from science as well as from the general public – offers insight into contemporary models of perception and experience. It ultimately acts as a metaphor for the nature of thought itself.

The phenomenon of olfaction encompasses two very different kinds of thinking. One is a primitive function rising through the reptile brain of raw physicality and passing into the emotional centers. This is the first part of smelling. The next part comes when the body-state is interpreted by the language centers of the brain, the very human cortical areas. Olfaction as a human phenomenon encompasses both semantic and episodic memory, as well as the conscious and unconscious mind. Granted, all sensory phenomena cross this threshold. Olfaction, however, is unique due to the way the information is translated.

The nature of olfactory input makes it fundamentally incompatible with language, and thus with the collective knowledge it serves; it is the epitome of "dirty data." This transition – from perception to description – is the focus of this text. Because olfaction exists at the nexus of the primitive and the intelligent, there can be neither a machine-readable language nor a universal language for smell. Consequently, smell can never truly become "clean data." This observation has implications for the role of scientific objectivity and for epistemology as the creation of knowledge structures.

The study of olfaction can help us achieve a deeper understanding of the knowledge graphs used in machine intelligence, to take just one example from the forefront of technological discovery. Here, in the world of olfaction, text is a limitation. In fact, language itself is a limitation. If we can safely assert that taxonomy, classification, or organization, in a most ideal form – i.e., machine-readable text – relies on a language, then the classification of smell falls outside this ideal form. It thus requires us to rethink the classical objective pursuit of epistemology as an ever-expanding sphere of interconnected, semantic-based knowledge.

I have been an art teacher. After 10 years, I have come to realize that it is more about teaching than it is about art. That means spending a lot of time considering how people think, how knowledge is created and accumulated, and especially about the barriers to knowledge construction. Language is one of those barriers. I teach visual arts, a subject where much of the knowledge evades language altogether because it comes in the form of a training of unconscious sensorimotor and perceptual processes. *Do you see that?* Terms like "seeing" are only one word for a whole suite of slightly different things. In our everyday lives, we need only to see. In contrast, to be an artist, one must perform edge definition, value calibration, proportion measurement, and so on, until the simple act of looking at a thing becomes a kind of math problem where the solution is the true visual nature of the object. Successfully executing this process requires constant training via heightened aesthetic sensitivity matched with a means for communicating the experience via objective scientific knowledge of the subject in the form of a carefully selected lexicon.

This linguistically mediated, disciplined sensory awareness naturally migrates to other modalities. This is what led me to begin my study of olfaction. Now the question became – *Do you smell that?* The problem arose when I found that I could not communicate properly with others about the things I smelled. We had no common language, because no common language exists. The problem got worse, however, when I discovered that objective scientific knowledge offered no respite either. If an artist wants to learn more about how vision works, then there are many places to turn for information. With smells, this is not the same. Smell is the least studied of all of our senses.

There are more books in the Library of Congress about perfume bottles than there are about smells. Of those listed, books on the sense of smell are overshadowed by three olfactory-related subjects: perfume, pheromones, and aromatherapy. These works constitute the majority of books available to the curious smell reader. There are only a handful of works that provide a serious treatment of our most hidden sense. The most important of these were written by pioneer olfactory psychologist Trygg Engen in the early 1980s. However, aside from his introductory work, and with the single exception of a collection of psycholinguistics essays titled *Speaking of*

Colors and Odors (see Dubois, 2007), none of these listed books does much to investigate the language of smell.

I mention this only as an indicator, one that reveals a part of our human experience that is very much absent not only from general public discourse but even from the upper realms of scientific endeavor. The engines of misinformation that distort our collective store of knowledge – pseudoscience, superstition, urban legends, and conspiracy theories, as well as paradoxes and illusions, both cognitive and sensory – all propagate through the field of olfaction uninterrupted by science. In our efforts to acquire knowledge, where do we turn? We cannot google smells. If we are living in an era named only one generation ago, the Information Age, then how is it possible that we know next-to-nothing about one of our only five senses? I find this astounding, and, as an educator, unacceptable.

This is what compelled me to write this book. In my search for this "language of smell," I read every book I could find (which was not much), as well as hundreds of scientific articles (which have a language all their own), in addition to thousands of threads from web-based forums. Of course I failed, because there is no such thing as a language of smell. The contents of this book are both a record of my failures and an attempt to map the perimeter of ignorance in the hope that the topic will be penetrated by someone with more ability than I claim to possess. This is a subject that is rarely addressed, and for that reason alone I hope its value can be assigned accordingly.

This book is separated into three parts. The arc of exposition follows from explanatory to exploratory, and, ultimately, to speculative application. The first section, Clarification, serves as a primer on the mechanics of olfaction, following the journey from evaporated organic molecule to conscious experience. It explains what smell is, where it comes from, how it works, what makes it different from other senses, and what its notable characteristics are. Because the olfactory phenomena are largely "below the radar" of conscious thought, it is important to clear up the ensuing misunderstanding that arises from ignorance. Because smell varies wildly regarding subjective experience, it is also important to address the biases and limitations that shape its effects within the individual.

The second section, Contradiction, gets closer to the heart of the matter. The "language of smell" serves as evidence of the disjointed transition from the lower to the upper brain. The book narrows its focus to explore olfaction as a linguistic phenomenon and as a subject for classification by exploring where the lexicon of smells comes from, how it has been organized in the past, and what the limitations of these methods of organization are. Because olfactory stimuli are mediated to our thinking selves via the unconscious limbic system, the resulting structures are highly disorganized. They therefore require a variety of strategies for organization, as opposed to a unifying rule.

Reversing the looking glass, Part Three: Confusion, probes the frontier of scientific knowledge as it expands into public discourse. Contemporary subjects of interest – artificial intelligence, multidimensionality, and quantum theory – are posited against the paradigm of olfactory experience as a vast information system at work. In this final section, the reader is asked to imagine a world where olfaction is fused with other puzzling topics in alchemical thought experiments meant to transubstantiate new knowledge. This speculative exercise is an attempt to bring the reader to novel discoveries of the hidden relationships that exist in the most unlikely of places.

In closing, a few explanatory notes. The first mention is about the footnotes and citations. An ancillary purpose of this text is to act as a sourcebook for the most widely recognized sources of olfactory study. To accomplish this goal I cite as many primary sources as I possibly can. Next, I concede that there are enough footnotes to warrant an entirely separate text. Another objective of this book is to ameliorate the confusion that surrounds olfactory study. So, if I err in my excess, it is on the side of caution. The last thing I want to do is to lead the reader to false assumptions based on incomplete information.

Next, a note on brain models. The brain is an extremely complex entity, and I am neither a neuroscientist nor a psychologist. In this text, I use two very basic models to talk about the brain. The first is the hemispheric model, or left-right model, which I have used for years to simplify visual-perception concepts to my students. It is understood that the science of the brain has since moved far beyond this simplistic, binary theory. However, the basic tenets used by Betty Edwards's *Drawing on the Right Side of the Brain*, for example, are still indispensable in the classroom. Likewise, and for what it's worth, much of my thinking-on-thinking is influenced by the bicameral mind theory of the behavioral psychologist Julian Jaynes, who points to hemispheric crosstalk as the origin of subjectively conscious, self-reflective thought. Jaynes's theory uses language as the vehicle for information traveling between hemispheres, making it relatable to olfactory experience as it is transformed by the brain from chemical stimulus to verbal description.

The other brain model I employ is the triune model of the reptile, mammal, and human brains, which are nested inside one another as a three-dimensional artifact of neurological evolution (see Figure 3, Chapter 2). This is the model that most frequently appears in the literature on Olfaction because the olfactory system resides in the mammal brain, and it is distinct from the human cortical areas – that is, until it becomes language. The relationships among the components of the triune model are essential for understanding olfaction. Hence, this model is an obvious choice for this text.

Finally, some thoughts regarding the influential template for information systems and data visualization. Although this text claims to put olfaction at the intersection of human experience and information science, and yet I am not an information scientist, I would like to provide the reader with the following schema. My thinking regarding olfactory experience as an information system stems from the Flavor Networks of Barabási Labs at the Center for Complex Nctwork Research at Northeastern University (Ahn, 2011). These networks are comprised of recipes, their ingredients, and the molecular constituents of those ingredients, which they call flavor compounds. The resulting network graph, a part of which is reproduced in Figure 1, reveals the relationships within cultural and regional cuisines as a form of data visualization made ubiquitous by the quintessential hub-and-spoke representation of a social network. In place of people grouped together by the friends they have in common, food ingredients are grouped together based on the flavor compounds they have in common.

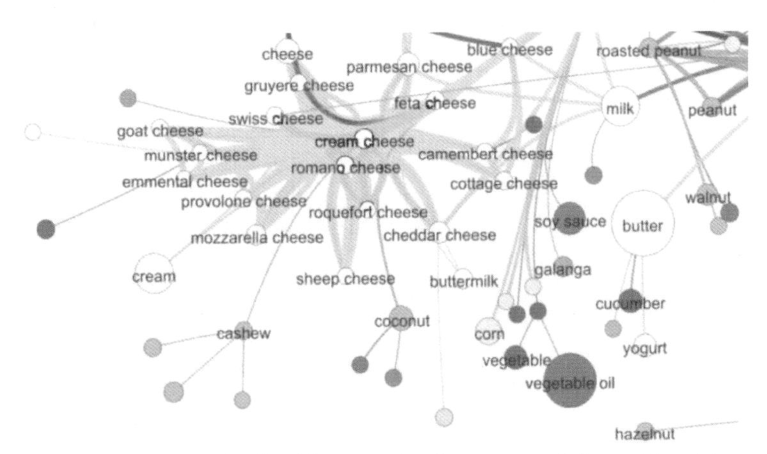

Figure 1: A portion of the "backbone of the flavor network," reproduced here courtesy of Barabási Labs (Ahn, 2011). This is a cropped portion of the full graph showing a grouping of cheese-related food ingredients. The cheeses are clustered together because they share many flavor compounds. Notice, however, that cashew is related to cheese (although most of the other nuts are isolated from the rest), and coconut is related to cheese and not to any fruits or vegetables (in fact, the mold that grows on coconuts is the same as that on Roquefort cheese).

This was the inspiration that led me to create my own version of this template, albeit for smells. Using visualization software and the Sigma Aldrich Catalog of Flavor and Fragrance Chemicals, I generated a networked graph of 1000 chemicals and their 3000 descriptors, all grouped together by their shared molecules. Alas, there are more “smells” than those used in the flavor and fragrance industry. Consequently, to compile a database of the magnitude required for a content-addressable smell network is, for lack of a better word, impossible. Nonetheless, what is left for the reader to imagine is a *hypothetical* smell network, similar to that described, only large enough to contain all smells as they occur to all people. That distance between the possible and the impossible, between the perimeter of ignorance and the advance of knowledge, is where we are headed in the next part of this text.

Introduction: The Internet is Anosmic

On April 1, 2013, the home-node of the internet, google.com, invited users to interact with their new smell messenger, Google Nose Beta. "The newest sensation in search" invited users to smell their search results in the "sharpest olfactory experience available." Of course it was a joke; it was April Fools' Day. But many of us might fail to recognize why this is meant to be funny. The internet is anosmic, that is to say, the internet cannot smell.

This simple premise – that the internet cannot smell – is the basis for this book. It may seem trivial, or absurd, but upon further inspection, it begs the question: In the Information Age, and the age of Big Data, how can one of our only five senses remain so cryptic? In pursuit of this question, this book explores olfaction as a phenomenon, and in particular as a linguistic phenomenon, that is at odds in every way with our contemporary model of information gathering and knowledge construction.

Structure of Knowledge, Perimeter of Ignorance

Where else should we look for the structures that hold our collective stores of knowledge but in the artifacts they have created? And where else could our attention be so captivated but at the limits of those structures, the horizon beyond which knowledge dissipates into the ether? The internet as a cultural artifact has become a kind of anthropic microscope, a metadata machine that calculates the latent desires, unconscious tendencies, and predictable patterns of its users' behaviors.[1] It records how we use it, and it thus avails us of a plethora of information about ourselves. Some of the most surprising observations emanating from this human-focused lens come not from the information accumulated therein but from those areas of knowledge that are left out.

For the subject of olfaction to be left out, for example – this is strange. This isn't flint-knapping, or the history of the ampersand. Rather, this is one of our sensory organs. Understandably, the internet cannot smell in the way it can see and hear. It is not a medium suited to chemical communication, so we should expect this. However, the sensory organs are the only mechanisms by which we can accumulate knowledge in the first place. They are not technological extensions: They are the technologies inside us. In this technophilic world, for one of them to contribute so little is suspicious.

---

[1] *To determine exactly whose culture it is falls beyond the scope of this text.*

Smell, both as a sense and as a subject of study, is an approximation. This inherent characteristic has kept it from the inner circles of knowledge, those places more commonly employed by the other senses. As a phenomenon, smell simply does not match our *modus operandi* for constructing and applying knowledge about a subject. It cannot be classified because there are simply too many smells. It cannot be verified because the words we use to describe it are subjective. It cannot even be unequivocally identified, because we its presence can be induced.

If we are truly living in the Information Age, brought upon us by the internet, then olfaction marks its boundaries. To venture beyond these limits requires a way of thinking that differs substantially from the one that brought us to this point. Olfactory study requires us to redefine the function and content of the language that we use to communicate, of dimensionality, of intelligence, and of metrics themselves.

Technology as Metaphor

Technology is often thought of as an extension of the human body and its sensory apparatus; for example, the catapult as a mechanical arm and the telescope as a magnificent eye. The computer is an extension of the mind itself, an exocortex. The internet is no different: It is an extension of our memory as a form of external storage. Writing itself is an external memory, and one of the first technologies. Along this line, even language is a technology that allows us to share memories within ourselves and amongst other humans.

Nonetheless, this anthropic memoryplex we call the internet is first and foremost a semantic web, and the most complex lexigraph ever imagined.[2] It is truly a *living* information system, representing the largest and most easily searchable repository of knowledge known to humankind.[3] However, the internet is primarily a textual phenomenon. Hyperlinks, computer code, and search results – all of these things rely on text; that is, machine-readable, written language. If human memory can be separated into semantic and episodic, then this internet must be analogous to only one part of our memory, that being the logical, discrete, alphabetically accessible semantic memory.

This is the fundamental problem with smell and the internet: Smells do not have names in themselves, and the words used to describe them are lexicographically complex and are not consensually recognized. In a text-based repository such as the internet, we can neither index nor query a thing that cannot be named. In turn, even if it were possible for us to make the object of our search precisely understood, the results of such a query

---

[2]*Then again, there must be speculative fiction writers who make their entire living off of this imaginative exercise.*

[3]*It is in good faith that the word "knowledge" is used here, in full consideration of the breeding ground of confusion that the internet also maintains.*

would be conflicting. Instead of converging on one answer, they would diverge into a constellation to be interpreted by the searching human and not processed by a "thoughtful" machine.

The Limbodic Frontier

There is no such thing as a language of smell; at least, not in the way we typically use language. Smell is a chemical communication, not a cognitive one. Smell borrows words from other senses (sweet, green), and it hides in the background amidst psychosomatic verbiage (deep, sharp). Omnisciently, it designates hedonic valence (disgusting, uplifting). Most importantly, this paralanguage of smell *approximates*. Smells are ethereal, ephemeral. They change from moment to moment, and especially from person to person. They evade specificity. When it comes to describing a smell, the fewer words we use, the less someone else understands the exact smell we are expressing. A cloud of finely nuanced ambiguity – that is what communicates smell best. Synaesthetic, poetic, essence. A crystallized knowledge it is not, nor can it be.

It is in this way that smell acts as a back door into the unconscious mind. In this it also shares, with the rest of our ignorance, in exposing the limits of our collective knowledge. This is the subliminal realm of our existence, beyond those limits marked by language (and essentially the reality that language creates). If reality is "illusion grounded in consensus," then we can only truly know that which can be shared (Grossen, 1989). Communication is our means for achieving this consensus. Our reality, just like our internet, is a textual phenomenon composed of things and ideas with names. Whatever resides beyond these limits must operate beneath the level of subjective consciousness.

This subliminal aspect of smell does not render it impotent. Just the opposite: It is invisible and unspoken. And yet, it controls our body *in spite of this*. Smell is a primitive actuator, an orchestrator of our emotions, a system administrator of our episodic memory, and a navigator of the contingency map through which we traverse. Smell is a wild animal, living in our civilized midst. It flitters about but never lands on the textual substrate of human experience. The language of smell is one that only the mind itself seems to understand. We, as rational beings, are left out of the conversation.

The internet and smell are inversions of each other in that smell is also a massive repository of information, probably moreso than most of us would imagine. Every smell to touch an individual's olfactory receptors instigates a pattern in his or her brain that encodes all physiological data, along with all spatiotemporal data (and even social data such as the people present). Significantly, this pattern is never forgotten for the lifetime of that individual. Note that prior to this encoding, the individual's brain matches the new smell against every previously recognized pattern, riding the hedonic topography of recorded smells along the associative network. Only if there is no match is the smell recorded. In total, the olfactory system is a content-addressable memory – the content being the limbic states of the body, and the means of address being organic molecules. The results of such an operation aim to generate knowledge about the environment.

This is not the kind of knowledge that is acquired by consensual recognition and pulled back and forth to be cross-cancelled and "refined" into truth. Smells encode the body-state of the subject in its entirety – heart rate, breath rate, skin conductance, how full your stomach is and what is in it, how much physical, mental, and emotional stress you are experiencing, all the things that *you* were too busy to think about while they were happening – all tied-up at one end by a particular configuration of aromatic molecules.

This is self-knowledge, and it cannot be accessed in the same way one uses the internet. The autobiographical data that are linked to the smells one has encountered cannot be recalled at will. There is no text-eating engine to facilitate our search, no episodic rolodex. Perhaps the act of conscious, intentionally manipulated memory developed at a time much later than olfaction and is therefore disjointed and interrupted in its common circuitry with language. Perhaps it is simply too much information to place on a single point, for the accidental or abusive activation of such a complex memory network would be overtaxing on an already overworked brain. For whatever reason, one cannot "search" one's own database of smellcentric data (at least not without "artificial" stimulation by "real," exogenous odors).

All this having been said, let us now imagine that the internet *could* smell. In other words, suppose we could use our olfactory-induced limbic memory network in the way we use the internet. An intranet, it would be more aptly called, to summon and stimulate a bodily state, emotions included, at will, and *ad infinitum*; to transport oneself years into the past, to places that no longer exist. It is interesting to note that, to date, virtual reality technology has a difficult time convincing the hippocampus of discrepancies between the virtual space being simulated and the real space where the person is located. As much as we might try to "be" somewhere else, there is a part of us that knows otherwise.

Suspension of disbelief? Please – let us not forget our imaginations. What if the internet was not a textual phenomenon? What if, instead of words, the internet was a world of volatile organic molecules? What if we could search this organic world with our bodies? What would we do with it? With ourselves?

But alas, the internet is anosmic, and for smell, we can only rely on fragrance artists and chance encounters to relive our past lives according to some chemical poem.

For now, let us put this question aside and lay down the basic groundwork for olfaction, how it works, and some of its most salient characteristics. Following this first section, the second section will get to the heart of the matter, or, to be more precise, how we know what smell is. How *do* we know what smell is? As with everything else, we use science. However, there is one very important thing about smell that prevents it from ever becoming a scientifically objective phenomenon. That thing is the language of smell. This will be the focus of Part Two of the book, which will explore why smell cannot be objectively verbalized. Because of its intimate integration with the limbic system and the emotional centers of the brain, smell must be subjective. This subjectivity denies any universal lexicon from which to build a structure of knowledge.

Part One – CLARIFICATION

The study of smell is beset by many struggles. The average person misunderstands from the very beginning the concise limits of smell, confusing it with physical effects of the trigeminal nerve, or attributing all of its credit to the sense of taste. In fact, only a small portion of what we "taste" is actually processed by the tongue and its neural correlates. This explains why most people are surprised at how much "taste" is lost if they hold their nose shut while eating.

As for the next point, any explanation of smell that does not mention pheromones is incomplete. Pheromones are related to sex. Sex sells, and selling something is in many ways antithetical to explaining it. To be informed by the commercial industry in general and by grocery store checkout-line magazines in particular is no way to learn about a thing. And yet, these sources contribute to the majority of the public's understanding about how smell works. Although pheromones are discussed in many of the books listed in the Library of Congress under the subject heading Smell, the subject is poorly represented relative to any of the other senses. Perhaps this is not a good enough indicator – instead consider how common it is to have a course in the public education system dealing with smell. There are visual arts classes, music classes, and even culinary arts classes, but nothing to educate the nose. Human pheromones are to this day not fully explained by the scientific community, which offers little in the way of clarity or correction.

Only after we redefine the parameters of the olfactory system can we begin to model its inner workings. Smell is a product of chemosensation. Its stimuli come not in the form of distantly originating rays of light or waves of sound-shaped air. Rather, they are generated by evaporating chemicals that must physically enter our bodies. The source and the stimulus are the same; they are not separated by an intervening medium. This reality affects the immediacy of olfactory impact. Locked away in our most primitive gray matter, deep inside, where it is literally difficult to access, this sense often lies beyond the grasp of rational thought simply by way of its neuroanatomical proximity.

Once inside, smell is the only sense to use the brain backwards (permitting a slight loosening of terminology). Olfactory perception is processed in the limbic system, whereas the other senses use their respective cortices. Vision is processed in its visual cortex, but the "olfactory cortex" is the limbic system itself. Many of these characteristics result from the evolutionary age of olfaction because it was the first sense to develop. On the Tree of Life, it coincides with the advent of vertebrates, although it should be noted that everything alive uses chemosensation; it is required for life.

Olfaction is one of the lower senses, and it operates unconsciously. Our osmic experiences are stored automatically, along with all of their corresponding physiological data. They are released unexpectedly, to pour forth the most convincing virtual representations. This autobiographical moment, the most commonly discussed feature of smell, acquires its strength from the mountains of data stored in the memory of an olfactory experience. Its tenacity, which works in tandem with its emotional power, comes from its being stored in the limbic system, which performs like a "read-only memory" and stores even the earliest experiences in high fidelity.

Just as the persistent, autonomic characteristics of smell give it such power, they also leave it with vulnerabilities. Because our olfactory system is always on, it suffers from fatigue. Consequently, over both short and long periods of time, our ability to smell a particular scent decreases. Also, as a lower sense, smell is highly susceptible to cognitive override. Verbal descriptions, visual information, and any other form of persuasion can alter olfactory perception and even cause extended somatic hallucinations.

This is not the most comprehensive survey of olfaction. However, it does summarize the primary misunderstandings that muddy the waters of the more intricate and nuanced discussion that will follow in Part Two.

1. What Smell Is Not

Three concepts obfuscate a clean delineation of the sense of smell: taste, the trigeminal sense, and pheromones. All three overlap with olfaction on a hypothetical sensory schematic. Therefore, a careful articulation of the boundaries of each one is required.

Taste

Taste is so different from smell that it is considered its own sense. Yet, most people misattribute the majority of what is considered taste-sensation: What we taste is more a function of olfaction than anything else. When we eat with a pinched nose, then taste is truly limited to the rather small inventory of what the taste buds translate as sweet, salty, sour, bitter, and savory (umami). Without smell, the glory of gustation would be to the soul what anesthesia is to the body. These five taste sensations can be more suitably classified as physical *feelings* of a tingling or puckered tongue, for example, and not as the *aromas* we normally associate with food and taste.

The Trigeminal Nerve

The trigeminal sense is not considered a sense *per se*, because it exists between taste and smell. Sensations such as the coolness of menthol or the heat of Tabasco are received neither by the taste buds nor by the olfactory receptors. Instead, they activate another component altogether, called the trigeminal nerve. The trigeminal nerve is spread throughout the face and head and is similar to the nerves that give our skin the ability to detect changes in temperature. People who suffer from loss of smell can still sense the coolness and heat described above.

Pheromones

Many years of concentrated scientific effort have given us a reasonably clear picture of how pheromones work in relation to olfaction. Put simply, they don't. Although pheromones are produced by humans and they may tend to appear in conjunction with other molecules that are perceptible to us, humans left the pheromonal sphere of influence many evolutionary increments ago. A middle ground has recently been established where pheromones are referred to as physiologically manipulative chemical signals or "modulatory pheromones" in order to emphasize that although they may not operate in humans as they do in other organisms, they still affect people in the ways that other odors do (Jacob et al., 2002).[4] Although humans may communicate reproductive information or simple fear by way of their body odors, the existing scholarship suggests that these are the same kinds of odors that transmit the signals of rotten or spoiled food, for example.

There are essentially two modes of olfaction: one kind, which handles pheromones, and the other kind, which humans would consider the only kind. Pheromones work via a kind of hardwiring, where the stimulated response is either involuntary or unconscious. The other kind of olfaction (for us it is the only kind) is softwiring, which means that the response to a stimulus can change over time and across individuals. One is "a synthetic, memory-based mode that rapidly learns to form perceptual odor objects from variable, novel patterns of input" (Wilson & Stevenson, 2006: 33).The other is a fixed, predetermined response that allows no room for discretion.

To support the occurrence of pheromone transmission, human-pheromone proponents frequently cite the vomeronasal organ. This is the organ that deals with the "other" sense of smell. Although humans still have this organ, it is a vestige of another phylogenetic era when less advanced methods of decision making were evident in the biosphere. The vomeronasal organ is like the appendix – as far as we know, it does not do anything for us.

And so, to summarize what smell is *not*: Although most of what we taste is smell, smell and taste are disentangled by way of the trigeminal nerve. And, pheromones are for insects, animals, and advertising, but not for human perception.

[4] *There is a preference for the body odors of people with different genetically determined immune systems (their major histocompatibility complex, or MHC); see Wedekind & Füri, 1997.*

2. What Smell Is

Humans have many methods for receiving information from their environment. Leaving skin sensation for another day, there are four other senses we are most familiar with: the two higher senses of vision and audition, and the two lower senses of taste and smell (formally called gustation and olfaction). These four senses can also be divided into the distal and the proximal, respectively.

Chemical Sense

The first separation to be made between olfaction and the other senses of vision and audition is that olfaction functions via chemical reaction as opposed to vibrating electromagnetic or mechanical sound waves.[5] This is problematic for the would-be thinker, conjuring analogies amongst the senses in an effort to clarify them. The proximal and the distal senses are almost nothing alike: neither in the way the information is encoded in higher-processing areas of the brain nor in the type of information received. In fact, the respective forms of the stimuli go so far as to fall into two different branches of science – visual and auditory stimuli are studied using physics, not chemistry.[6] Being that all stimuli are reduced to some invocation of chemical reaction upon reception, however, the chemical senses are one step closer to the body's indigenous realm of interaction, where physics is ethereal and chemistry visceral.

As will be presented at greater length below, the chemical behavior of smell espouses its own variation on the *name-thing* problem in that when we smell a thing, it is the *thing itself* that impinges on our sensory receptors, entering the nose, crossing the threshold of the body's envelope. There is no interstitial space, for the body itself is the receiving medium. For visual stimuli and sounds alike, what we interface are disturbances in the radiative fields surrounding our bodies: The *source* of the stimulus is *out there* and not wiggling into our neurons.[7] (This is why there exists the

---

[5]*There is a continuing debate as to whether olfaction is a process of detecting molecular vibrations as opposed to molecular shapes, which is the dominant theory. According to Luca Turin in his 2006 book, it doesn't matter that the debate is still in play; his ability to use vibrational theory to synthesize odor molecules at a rate faster than those using shape-theory makes the point moot.*

[6]*The translation from stimulus to perception in both vision and audition is a chemically influenced process, but the study of the stimulus itself is conducted via physics.*

[7]*The olfactory sensory receptors are in fact* exposed neurons *(located outside the blood barrier) – the retina is encased in the eyeball, and the cochlea is on the other side of the*

sensory distinction of distal/proximal apparatuses: The latter detects information of things outside the body, whereas the former detects things already in contact with it.) A note should also be made here as to the measurable characteristics of any stimuli. Smell, sound, and light can all be measured quantitatively; that is, by intensity. Stanley Smith Stevens, American psychologist/psychophysicist, has termed this the prothetic quantity. Though smell does not, in fact, have its decibels or lumens, it is not this prothetic attribute that makes smell what it is. Rather, its identity arises from its metathetic qualities, those that define the *kind* of perception in the way light has colors and sounds have notes.[8] In fact, "Perhaps the best way to characterize [smell] is to say that it primarily transmits information of qualitative similarity of percepts rather than psychophysical difference in intensity" (Stevens, quoted in Engen, 1991: 123).

The chemical realm also requires a kind of virtuosity in the way information is processed. As a stimulus, chemicals can take almost any form imaginable. Compared to the one-dimensionality of scaled frequency reception, the sensation of physical molecules is a "three-dimensional" affair. There are 450 receptors that somehow make sense out of the veritably inexhaustible list of odor chemicals. This complexity is in stark contrast to the relative simplicity of the visible light spectrum of vibrational frequencies, which are detected with only three kinds of photoreceptors. Colors exist on a continuous spectrum, meaning that adjacent colors look similar and therefore can be grouped together. Although the concept of a color spectrum for visible light is a relatively recent development in history, it seems most intuitive in its organization. Any color sensation can, with tremendous accuracy, be placed on a spectrum of seemingly infinite resolution and expressed in terms of very distinct values. Color variations are so accurate, in fact, that they are used to judge the speeds and distances of objects at the extreme edge of the known universe.[9] In contrast, smell, as a chemical sense, gives us only one piece of information – good/bad. There is no spectrum of variation. When viewed from this perspective, the other senses are like analog, whereas smell is digital.

Although this seems a paltry output for a sense that enlists 2% of our genome, the significance of olfaction can be better understood in articulating the phylogenetic development of our organism, for it is under this next subject that the criticality of smell can be reckoned with.[10]

---

eardrum.

[8]*Note, regarding the complications inherent to using intensity to characterize smells, grapefruit or civet can be pleasurable at low concentrations, but they become nauseating and disgusting at higher intensities.*

[9]*In fact, in a visual* coup d'état *over olfaction, toxic vapors can be detected by their unique photophysical properties (Saha et al., 2013).*

[10]*This is an order of magnitude greater than any other system in our organism; smell has them in the hundreds, others in the tens Each olfactory receptor type has a gene coding for it, making roughly 450. There are about 25,000 genes in the human genome, classified into*

Phylogenetic Development

"Olfaction is the first sense"
-Wilson & Stevenson, *Learning to Smell*, 2006

One very pertinent concept to this discussion on sense, as it pertains to olfaction in particular, is that of the positioning of phylogenetic development. The human organism has been assembled over a long timespan and as part of a lineage that reaches back to the first organism. On top of the morphological development of the organism is a layer of neurological development wherein the workings of the body and its sensory systems are configured and reconfigured in relation to the perceptual and cognitive processes required for the ever-changing body in relation to its environment.

Some of our neurological modules were assembled during overlapping timespans and thus became interrelated. When things are positioned near one another on such a trajectory, they can be referred to as moving in parallel, in regards to their respective neurological stewards. Note, for example, the adjacent neuroanatomical positions of the somatosensory modules: the brain-areas for sensing one's fingers, hands, arms, etc. are located near one another in the physical anatomy of the brain in a place called the somatosensory homunculus (see Figure 2). This chronological overlap creates a synergy from which novel interworkings develop. The parallel appearance of bipedalism and trichromatic vision is an example of this confluence, and one that is suspiciously "responsible" for the apparent demotion of smell to a secondary status. A nose close to the ground is a good sense organ. However, at five feet in the air it begins to lose its return on investment. Being this far off the ground, however, confers an advantage on the distal sense of binocular vision. What we are witnessing here is a redistribution of resources within the developing organism by natural selection; trichromatic vision seems inevitable, when paired with bipedalism.

---

about 10,000 gene families.

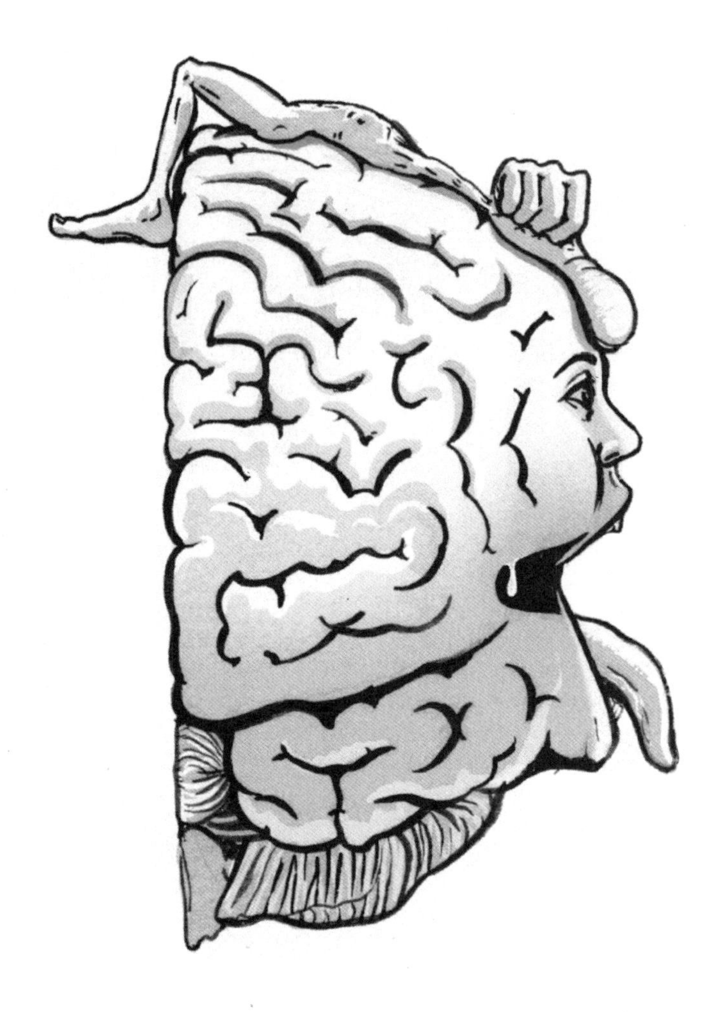

Figure 2: Somatosensory homunculus represented as its related body parts, based on the MacLean neurological model. *Illustration by Joe Scordo.*

Olfaction as a sensing mechanism was in use long before humans evolved, appearing first in fish: "It is a sense more suited to a fluid medium" (Morris, 1984). Any development that takes place far after the advent of olfaction will have to be retrofitted into the existing framework. Language, for example, has a most peculiar neurological relationship with olfaction. All of the modules that were already in effect in the organism at this point in its evolutionary development had more time to work together and tune a kind of resonance with one another. Furthermore, during this time the brain itself was practically nothing compared to the one now reading this text.

When discussing the sense of smell, the subject of phylogenetic development is a reference point used to estimate the amount of knowledge required to understand how any two modules interact. Olfaction and language are worlds apart in this regard, and a labyrinth lies in between.

Neural Phenomenon

“Sections” of the brain are clusters of similar neurons that are connected to other clusters via their axons. They are, in physical terms, too complex in their arrangement to comprehend spatially, so we delimit them to some basic “sections” (Sapolsky, 2010).

Concerning the history of the brain, there are two areas of interest: These are the “nose-brain” and the full human brain. Literally encapsulated in higher-processing cortical mass and symbolically shrouded in millennia of phylogenetic adaptation, the “nose-brain” is considered the ancient core of our brain (sometimes called the paleocortex), and it can be seen as a separate system from the rest. That it is deep within the mass of the brain, literally hard to get to, is (relatively) vaguely articulated in all areas of study, science included.

This part of our neural equipment is more officially called the rhinecephalon, but it might also be referred to as the animal brain, or, more specifically, the mammal brain, as to distance it further from the reptile brain (see Figure 3). The term itself is revealing of the experiments conducted on rats at the time – their olfactory bulb was observed to take up a large segment of their brain mass. Semantics aside, the rhinecephalon is composed of very primitive brain parts, such as those found in reptilian or earlier mammalian creatures.

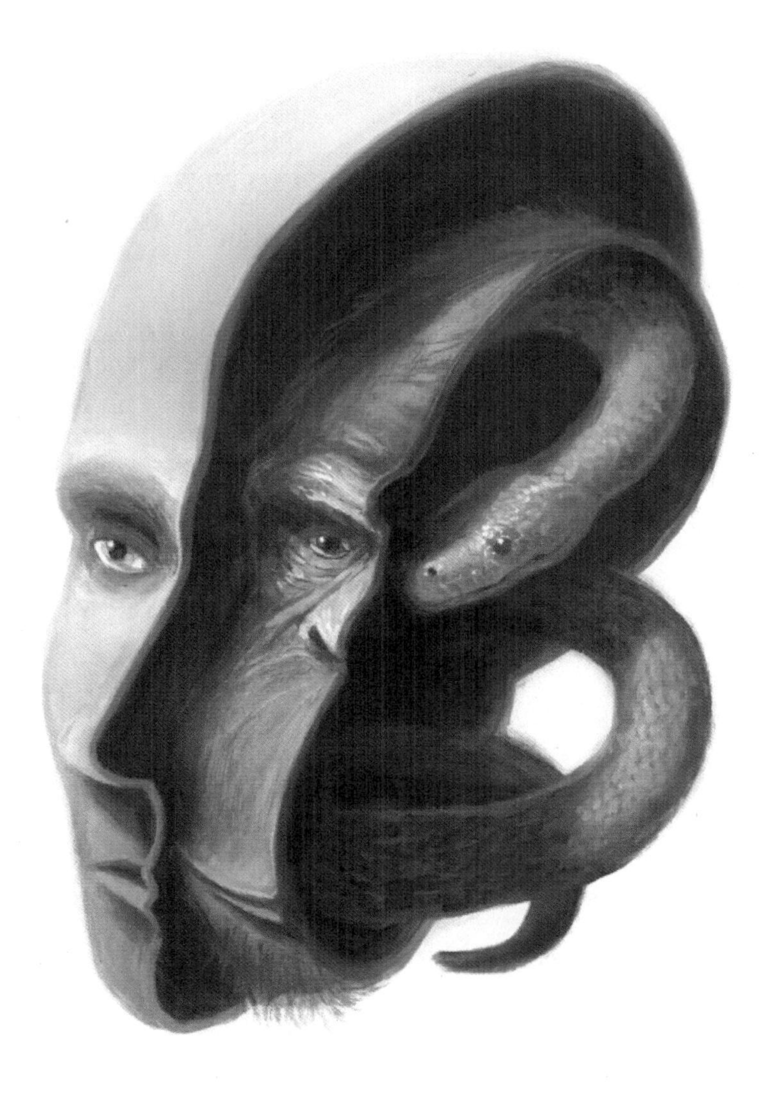

Figure 3: This concept of the reptile-mammal-human brain, or "triune brain," was first postulated by the neuroscientist/psychiatrist Paul MacLean in the 1950s. *Illustration by Joe Scordo.*

All other parts of the human brain as we now know it are, for the purpose of the discussion herein, separate. Again, as noted, all of these other brain parts have had to be, in a sense, retrofitted into the nose-brain. For this reason they make for a very complicated schematic of overall perceptual and cognitive processes. As an introduction, only the parts limited to olfaction will be presented. Later on, it will be important to elucidate the ways this nose-brain interacts with the other modules such as vision and language.

Entry: Receptor Neurons

The neuroanatomy of our sense of smell begins in the nose, where neurons have found a way outside the body's blood-brain barrier, and they swim in a layer of mucus on the surface of our epithelium. These cells are called the olfactory receptor neurons. They take information from the chemicals that enter your nose, either from the air around you or from the evaporating molecules of the food in your mouth (which are pumped up the back of the nasal passage in the act of chewing). The odor chemicals are encoded by one, or many, of the hundreds of receptor-types available (see Figure 4). The actual mechanism by which chemicals are turned into electrical information at the receptor stage is missing a complete explanation. However, we at least can think of these sensory neurons as converting either a chemical's structure or its vibrational frequency into electrical information that can then be transmitted throughout the rest of the brain.

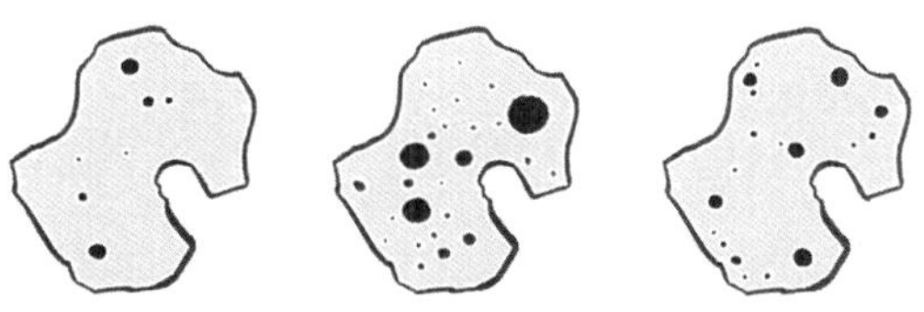

Figure 4: Differential responses of populations of olfactory sensory neurons to three different types of odors (oranges, banana, and paint thinner). *Image found in Davis and Eichenbaum (1991: 16), reproduced by Joe Scordo.*

More misunderstood than this mechanism is the way a few hundred types of neurons render the thousands of smells we perceive. Making sense of this broad spectrum of overlapping individual neurons appears to be a job better suited to a mathematician or an information scientist than to a chemist or a neurologist.

First Interchange: The Limbic System

The sensory neurons and the information they have detected terminate at the olfactory bulb, where information is then sent through the limbic system. This part of the brain gets its name from its relation to the limbs of the body, the soma, which it controls via parts like the cerebellum, but – most importantly regarding olfaction – the amygdala. The amygdala is the primary component of the limbic system that works with our olfactory sense. Significantly, the amygdala also is in charge of emotional regulation. Consequently, our sense of smell is very closely tied to our primitive, animal, nonthinking selves. From the animal's point of view, there is no distinction between olfaction and emotion. Rather, the animal's survival program reads them both the same.

The process of olfaction is intimately integrated with the limbic system, thus placing this less-known sense in a central position regarding the ultimate control of the physical body and its more mechanical operations. The anatomical proximity of the olfactory neurons within the limbic system is also indicative of the logistical priority of our sense of smell, both in the time it takes to register and in the load of emotional/physiological modules it can affect. Note here, this talk of proximity refers more to the number of synaptic interchanges along the wiring (the axons) than to the length of axonal distance between sections. Equally important, it is not a reference to the spatial adjacency. Rather, it is more about temporal than spatial proximity; hence, it is a measure of Hamming distance. Due to the seemingly unintentional consequences of evolution and its eventual exploitation of embryonic development, a critical space-time differential is staged between the developed and the developing regions of the brain. This critical differential, for example, sends the amygdala's connection to the hypothalamus all the way around the perimeter of the limbic system, instead of using a much shorter distance, which is perfectly viable, given that the two sections are situated next to each other. This makes no sense, unless under the rule of conservation of synaptic interchanges, where fewer interchanges result in a more efficient system.

Regarding the purpose of odor perception, there is no time for contemplation of an odor stimulus. Upon smelling rotten milk, your head turns away, involuntarily. The signal is not distributed to other parts of the brain to parse and deliberate. The first creatures to use olfaction did not have the luxury of contemplation – they had no "contemplating apparatus." Most mechanisms of the olfactory system do not include contemplation because there was no such thing during its evolution. Consensus, as the end-result of deliberation, is not present in the limbic system. It is activate-inhibit, and that is all.

One last point, while on the subject of the limbic system: Olfaction is, through the act of breathing, directly linked to the involuntary nervous system. It cannot be turned off. We can close our eyes, and stuff our ears, but pinch the nose… (Engen, 1991).

Next and Last Interchange: Olfactory Cortex

In neurological parlance, a cortex is a place in the brain where information integration, or meaning making, takes place. Therefore, what we typically refer to as sensation is probably more accurately described as reception-perception. Reception is the transduction of stimuli into brain-ready electrochemical signals, and perception is the convergence of these signals into meaning after they have been widely distributed across various neural modules. In other words, when a receptor is activated, it creates meaning for the *brain*, and the cortices create meaning for the brain *user*. One informs the detection device, the other the device controller.

Every sense has its own cortex, the place where raw, "meaningless" sensory information becomes meaningful perception. For vision, audition, and gustation, the route of transmission is a direct path from the receptor to the thalamus and then to the respective cortex. Smell is different. Its sensory input is relayed directly to its cortex and not initially to the thalamus. Therefore, the olfactory cortex is backwards as far as typical cortical behavior is concerned. Olfactory information is delivered to other parts of the brain independently of its cortex. Whereas for the other senses information is processed in the cortex, the olfactory cortex receives preprocessed information. To call this information *preprocessed* is a slight misnomer, because the "preprocessing" system – in this case, the limbic system – does not process, or integrate, anything. It is too primitive of a system. When compared to later neural developments, whatever the limbic system does, it is not processing. In this way, the olfactory cortex works in reverse. The further implication of this reversal is that the olfactory cortex is incidentally an area of convergence for somatosensation and emotion. In light of this overall "cortical inversion," the olfactory system truly is a brain within a brain.

This observation can be related to the previously mentioned concept of the positioning of phylogenetic development: The olfactory cortex was developed prior to the "practice" of using the thalamus to direct sensory traffic, because the other senses (and hence their corresponding cortices) did not exist at the time. Put simply, there was no traffic. In fact, at the time when olfaction was a kind of advanced technology, the schematic was so relatively simple that only two synaptic interchanges were required for odor perception: one between the bulb and the limbic system, and the other from the limbic system to the cortex. Any other sense uses up to five interchanges to transmit signals, making for a much more indirect route. In so many ways, smell is the "gateway to the mind" (Engen, 1991: xiii).

3. Strength

Nothing can unleash, unbidden, the torrent of emotion that entirely subsumes your mind and your body, pausing your beating heart, squeezing and sucking the memories of a lifetime through a hole the size of a single instant, leaving you helpless, dominated, and exposed; nothing can do this to you, but smell.[11] The reader will be spared (for the moment at least) the most famous chunk of text in all of smell-mentioning literature, that of Marcel Proust's Madeleine cookie, evaporating his stream of consciousness only to condense once again into the hermetic capsule of nostalgia. The reader knows, after all, exactly what has been described here because the reader is human, and there is no human who has not experienced this single yet uncontestable power of smell.[12]

Credibility verified: There are two words that can be used to describe that which makes smell-memory so powerful – tenacity and potency. Smell-memory is tenacious in its duration – it lasts forever. Once a smell-memory is encoded, it remains for the life of the organism. It is also tenacious in its resistance to revision. Other memories can be changed. In fact, our memory is constantly rewriting itself based on new and potentially contradictory information. Smell is potent in that it evokes *emotional* memory – not dates and names and facts, but *personal*, loaded triggers directly linked to our hormonal secretions, heart rate, and other physiological processes. Also, smell is potent because the details of the memory it activates are so comprehensive. The wealth of details delivered by a strong smell-memory is impressive, to say the least. Granted, this wealth of details is made more impressive in light of the age of the memory within the organism. Most memories are subject to revision and decay. In fact, a preexisting condition of memory is that the older it gets, the fewer details it avails. By contrast, smell-memory, especially the powerful kind that stops us dead in our tracks, is likely to come from a time early in the life of the organism. By comparison with any other sense-memory, it will *seem* exceptionally prodigious. However, the magical superpower of smell is not merely a matter of relativistic recalibration. Rather, for reasons that will be expanded upon in this section, our sense of smell stands alone from all others. By picking apart the pieces, we can establish a workable understanding not only of how memory itself functions, but also how this otherwise underexplained and universally experienced Proustian phenomenon comes to be and how and why it moves us so dramatically.

---

[11] *This also serves as an adequate description of the effects of trauma; refer to work by Kline & Rausch (1985) and McCaffrey et al. (1993) which looks at the ability of odors to instigate traumatic events related to war long after it is over.*

[12] *There are however the conditions of general anosmia, or loss of the ability to smell, and congenital anosmia as the loss of smell since birth.*

Memory and experience exist at either end of a static-dynamic spectrum. Along this spectrum, the position at which a particular moment resides is a function of its newness. When we encounter something completely new, we have no memory to work from. Consequently, we can be said to experience that moment in full, present awareness. Having encountered a similar situation, however, we rely on memory to translate, and, in a sense, to *substitute* for that experience. This gives rise to the term "autopilot,"and it provides the existential assault fueled by commuter traffic: When doing the exact same thing every day, such as commuting to work via a specific route, we are not experiencing the event firsthand, as it were, but through the lens of memory. In this analogy, memory is the static version of experience. It does not change; it is the .jpg, not the .mpg. Experience is then a sequence of these static "images" superimposed on the dynamic milieu. Experience "in the raw," a hypothetical concept, is much too bombastic in its detail to process all at once.

In a 1989 essay on vision, sensory scientist Richard Held stated that our perception is "reflective activity rather than passive reception." Essentially, a stimulus not only imprints itself on the blank slate of our receptors, but it is then manipulated by our memory of the stimulus, in concert with all of the memories of the co-occurring stimuli, especially including information about the "perceiving self and the general context and not merely the stimulus input" alone (139).

The perception of an experience, then, falls somewhere between the presently occurring experience and the individual's memory of a similar experience, via this dual-processing system. This model might be more readily schematized by conjuring a metabolic limit on the neural-cognitive processing of "new" information. Put simply, it is more efficient to *guess at* what a stimulus *probably* is than to fully analyze, synthesize, and evaluate it anew *every time* we encounter it. Perceiving a new stimulus – a face, for example – requires a multitude of processes, such as comparison to all other faces in memory, assigning and disentangling corresponding semantic information (that is, the "name" of that face) and even facial expressions and their meaning within a given context. To simply match the face with one from memory and then generate the information from *within* the individual, rather than from the stimulus/context itself, is a more efficient process. Metabolic resources have been saved for another moment.

To move forward in the discussion of how smell-memory works, we must accept, or at least imagine, that every moment is not actually being experienced in its fullness. Instead, much of the work of experiential processing is offloaded to the memory, a kind of virtual experience that is simulated using a store of all of one's previous perceptions. The result –

object recognition, for example – is still the same. Only the means have changed.

Potency

The nostalgic force of smell-memory – the Proustian moment, as it is called – is activated or simulated by a massive set of stored experience. This set of experiences is the episodic memory, sometimes called autobiographical memory. Its inventory is made up of emotional body states. Let's take a look around.

The Episodic Memory

Experience, whether being faced for the first time or as a composite of new perceptions and old memories, is subsequently encoded into the storage depot of memory in one of two ways: episodically or semantically. Any stimulus that is received for the first time is written into memory in one of these two ways.[13]

Semantic memory is the keymaster to the language-based partition of memory, the thing that travels backwards from visual images of words, to sounds of words, to the voices themselves – on one end – and the stimulus being referred to on the other. The semantic label precedes access to the memory of the thing. Our experiences exist in between these opposite ends of this memory-retrieval tether. In another partition of our memory, the stimulus is still on the latter end of the tether. On the former end is *every other memory* associated with the stimulus in question, at the time of initial entry. This is the episodic memory, and it is the inversion of semantic memory.

Whereas semantic memory is a tether between label and a memory, episodic memory is a multitude of tethers between the memory and *all* of its episodically related memories. Trygg Engen in his book on the psychology of olfaction argues that episodic memory has "no identifiable attributes of its own but exists as an inherent part of a unitary, holistic perceptual event" (1991: 7). This episodic macromemory is vividly revealed in the Proustian phenomenon referred to above, where a single stimulus (the Madeleine cookie) retrieves an entire simulated world of the past.

---

[13] *For practical purposes, both ways of writing stimuli into memory are at work at all times, presumably from the moment the first stimuli are received. The concepts of phylological and ontological memory complicate this, however, and will be explored later.*

Although these two areas of memory have been herein referred to as "partitions," they are not physically separated locations or even categorically distinct entities. Rather, their partition is one of behavior, interactivity, or functionality. One kind of memory writing yields one type of connectedness amongst memories, and vice versa. Both types utilize the same base of received stimuli, and these stimuli in actuality are being written-in using both types.[14]

Thinking of memory in terms of this dichotomy coordinates various other opposing sets with which the power of smell-memory can be better articulated. Olfactory perception rarely bifurcates during its encoding process to be semantically indexed (with the exception of those whose professions depend on it).[15] Olfaction does not use semantic labels during perception. It is a primitive sense: Its golden age flourished at a time when language did not exist. Therefore, olfaction does not rely on and does not contribute to the interworkings of our semantic memory.[16]

Cognition supreme, human experience is mediated by the semantic memory, not the episodic. The labels we apply to semantic memory – *indexed, categorized, labeled, discrete* – are similar to the labels we apply memory itself. *Declarative memory* it is called, in opposition to the non-declarative. Episodic memory is not organized in this way, however. It is gestalt, multisensory, richly detailed, and autobiographical. It records, in one memory-unit, the entirety of an episode – spatial, temporal, interoceptive, and exteroceptive information. It records subject-centric information, like the emotions of the receiver, as well as the objective information of the source of the sensory stimuli, or any and all other co-occurring external stimuli, for that matter. These are called the autocentric and allocentric aspects, respectively. This multitudinous repertoire is invisible to the conscious mind, if only by way of its seemingly immeasurable volume of memory pieces. In the same way we do not normally consider the Earth as a separate and distinct unit of information in our description of the setting of a particular experience, so it is with the episodic memory. It *is* the experience, as the Earth *is* the setting. It contains within it everything already, so why separate it and call it something else?

It is the observer, the subject, the receiver that changes the relationship here. The person who receives a stimulus has his or her interior world with which the stimulus must first interact and exchange before it can be called

---

[14]*Semantic memory, however, is dependent upon the receiver's language capabilities; in people who have not learned a language yet, their semantic memory is obviously less active. And to pull the string one notch tighter – things like gestures are a form of semanticity in that they are a proto-language, taking us further into the realm of the phylological memory and biosemiosis.*

[15]*Fragrance artists, for example, actually smell in terms of words; it is part of their training.*

[16]*Obviously, the truth of the matter is always more complex. Surely olfaction has its place in such operations, but that place is precipitous, and unfamiliar; the interactions between olfaction and semanticity will be explored in-depth to follow.*

into existence as a perception. Despite our being conscious – it could be called instead a *capacity* for consciousness – there are many experiences during which the constituent information *could* be consciously recognized. However, this does not happen. The entirety of episodic memory is encoded unconsciously; this is inherent to its non-declarative nature. Language and its explicit classificatory cognitive behaviors – the allometric identifications of semantic memory – have more in common with our working understanding of memory and perception. By contrast, episodic memory is a phantom, behind the curtain, unbidden, automatic, always on, and generating output suddenly and without solicitation.

Emotion

The presence of emotion in the arena of smell-memory is quite obvious. From what has been described thus far it should be clear as to why this is so. Emotion is a primitive system, just like smell. In fact, it is at times difficult or impossible to separate smell and emotion as they function within the organism. Emotion is a decision-making apparatus, precognitive, that weighs the potentiality of outcomes along various scales of interoceptive, physiological gradients. The internal, physical, bodily memory-record of a presented milieu is contrasted with the actual incoming information; that is, exteroception. The associated elicited memory combined with the thing itself, in conjunction with a desired state of homeostasis, activates a "response policy" directing the organism towards that desired state, and at a rate determined by the proportion of intervals in this comparison (Damasio, 2010). Again, this is all precognitive. Stimulus, physiological response, resulting interoception, autocentric memory superimposing, and final responsive outcome are all automated. To simplify this model, let us take it to the animal world, to our friendly synanthrope the deer. The deer relies on a phylogenetic store of information, combined to a much lesser extent with an ontological store, and all of it, regardless, containing biological information *only*. If information is not decoupled, abstracted from the body itself, then it is not cognitive. Pheromone-directed behavior, as mentioned previously, is not intercepted by thoughts of what the effects of pheromonic molecules *mean* to the body. The deer does not direct the behavior; the pheromones are the captain of this ship. There is no executive override. The body reigns, cognitive deliberation is a yet-mute sense, and thus a "primitive" system is described.

This is where our emotions live, and this is why assessing them, intentionally accessing them, and ultimately reckoning with them is such a difficult task. "An odor is a particularly unambiguous and unforgettable signal serving a primitive survival function" (Engen, 1991: 119). Assessment and intentional mental grasping are highly cognitive, conscious acts. These are two species of mental activity very different from each other, distanced by millennia of development. Imagine the sour-faced elder, shaking his head in confusion at the "youth of today," never able to fully decipher their dialect. Something is lost in translation.

This link between the episodic memory and emotions may account for the emotional potency of smell-memories, but it fails to explain their tenacity. Why do they endure for so long and with such fidelity against subsequent revision?

Tenacity

"Perfume is liquid memory."
-Diane Ackerman, *A Natural History of the Senses*, 1990

Most memories change and degrade, but smell-memories last forever. What makes them so tenacious is a confluence of factors. Some of these factors regard the nature of the stimulus itself; others result from the workings of memory.

By the Numbers

Steeled against the kaleidoscopic assailment of sensory stimuli and their subsequent memory modulation, the fortress of smell-memory is defended by many means, most of which could be explained by probability alone. Let us start at the logical beginning, by looking at the numbers. Two number sets are relevant here – one is simply the set of total smell-stimuli, and the other is the set of episodic smell-memories; that is, the amount of potential combinations between the available smell-stimuli and their corresponding stimuli-context.

Rachel Hertz, in her book on scent, reminds us: "How often is a deeply significant memory revived by the smell of coffee?" (2007: 74). Coffee is not the kind of smell in question here, but rather the "Madeleine cookie dipped in lime flower tea." Coffee carries with it the multitude of contexts in which it is encountered, thereby diluting its correlating episodic memory complex, and with it the intensity and the purity of its subjective emotional impact. It may elicit stronger memories of awakening, for example, but it does not activate a single, piercing moment of emotional and autobiographical clarity. In fact, coffee as a smell is more like color. We smell it in such multitudinous contexts, and so frequently, that its ability to stimulate our limbic system has been fatigued. Surely this fatigue is a key concept in the function of olfactory memory. But not all smells are like coffee.

There are more than 100,000 smells. No – there is an infinite number of smells.[17] People who actually need to make such estimates (scientists and fragrance industry professionals) put the number at 100,000. Most people have only about 10,000 in their "vocabulary," with the former-referenced professionals reaching into the 40,000s. In disambiguation, any particular smell is actually a composition of chemical compounds, each with its own individual smells. When these compounds are combined, however, they create wholly distinct odor objects. Peppermint and spearmint may smell similar (like "mint"), but they are in fact different. Consequently, when they are paired via episodic memory correlates, they can elicit different emotional states.[18] Straightaway, if a stimulus cannot be reduced beyond a base set of infinity, then it is less likely that an individual will encounter the same stimulus twice. Thus, the initial smell-memory is less likely to be re-written; that is, it appears more tenacious.

A particular brand of dryer sheets (also known as fabric softener sheets), however, is probably occurring in the same context every time it is perceived. The smell of dryer sheets in general will not occur at school, on the playground, at a restaurant, or at work. Further, the chances that a particular brand of dryer sheet will be encountered in various different locations are even less probable. Because smell is a *chemical* sense, it is inextricably linked with its *source*. And, by extension of its being a proximal sense, it is more closely linked with its *context*, thus creating the tightly knit knotwork of olfactive stimulus, source, context, perception, and memory.

Measuring a bit more carefully, still by the numbers, the human nose has about 450 receptors that, again, can combine with one another to generate, again, practically innumerable neural-coded odor objects. Color vision, by contrast, uses three receptors, which combine to create color stimuli that exist adjacent to one another in ways that are both objective and subjective. That is, if we were to measure colors in terms of frequency, yellow's measurement would be "close to" that of orange. By contrast, if we were to measure the colors in terms of a subjective response gauged by, let us say, the changing hues of a sunset, then the participants would tend to refer to yellow as the one that happens "before" orange. The base set of colors, unlike smells, can be reduced to a divided seven-colored spectrum, because the spectrum is the only practical way of categorizing electromagnetic radiation, and perhaps because seven is the number of concepts the human mind can grasp simultaneously.[19,20] The full concept of the spectrum

---

[17]*Chemical compounds are, for all practical purposes, unlimited in their number of possible combinations. (See also: Humans Can Discriminate More than 1 Trillion Olfactory Stimuli, Bushdid et al., 2014.)*

[18]*Examples of the differences in chemical constituents of odors can be found in books that use gas chromatography, amongst other techniques, to determine discrete chemical properties of odor-producing substances. See: Essential Oils Analysis by Capillary Gas Chromatography and Carbon-13 NMR Spectroscopy, Kubeczka & Formácek, 2002.*

[19]*See: The Magical Number Seven, Plus or Minus Two: Some Limits on our Capacity for*

requires that the subject maintain a mental image of the spectrum *in its entirety*: A spectrum of even 100 distinct units of division would defeat the purpose of the spectrum in the first place.[21] Regardless, the difference in the structure of the stimulus-complex of smell is vastly at odds with that of vision, making any discussion on the tenacity of smell grossly distorted against a visually normalized baseline.

So, via the lens of probability, we can compare both the sensory input units themselves and the way they are stored to imagine a diverse population of olfactory associations. These associations are recurrently reinforced over the life of the organism to produce a memory-patinated perception, originating from the very first experience with that smell, and changing very little thereafter. This is only an initial, rather casual observation, however, because smell and everything related to it are anything but logical or casual, as we would expect for anything tied so tightly to the limbic system of panic and emotion.

Age and the Unconscious

Next up, still looking at the numbers, is a comparison of the amount of episodic memory entities versus those of the semantic type, as a function of age. This is an ontological perspective – focusing on the development of the individual, not the species. Over the life of the individual there is great difference in the way memory works, and, by direct association, the individual's sense of smell. The olfactory sense, along with all others, diminishes with aging, but this is not part of the examination at hand. "Odor preferences are absent at birth and acquired with age" (Engen, 1982: 131). In this case it is not the olfactory sense *per se*, but the way memory is used during the different parts of one's life. As mentioned already, the use of semantic memory is dependent on the subject's aptness for language and semiotic engagement. Recollecting this fact, the young person's reliance on episodic memory is apparent and critical. We can imagine a chart where the two lines crisscross: Semantic memories start low in number and rise as

---

Processing Information, George A. Miller (1956).

[20]*Without treading too far off course, there is a general understanding that the color spectrum has historically increased in number, starting with binary division (which still may only be seen as a rod-phenomenon), on to the triad which includes red, (and since the binary opposition are not colors* per se, *this makes red the first "color") and differentiating from there. Western tonality is thought to have started at the separation of the octave, which is not an identification of a tone* per se, *because there are no other tones for it to relate to. Upon further division of the space* between *the octaves, tones begin to form. It seems that the further back we go as a species, the smaller the spectrum becomes, something that may look reversed on a similar plotting for the "vocabulary" of smell. See: Sachs 1943, Berlin & Kay 1969, Kay & McDaniel 1978, Deutscher 2011, Gladstone 1858, and Durham 1991.*

[21]*This section seems at odds with a previous one, on the smell as chemical sense, where it is compared with color, which "can be used to measure light at the far edge of the universe." In this case, it is a semantic or subjective issue that is generating the description. Surely, when the (only) seven colors of the rainbow are put to uses far-beyond-useful to the immediate human body and the basic human experience, they are actually not seven colors but an infinitely valued gradation, both measured and "seen" by a technological extension of the body, but not by the body itself.*

the individual ages, while episodic memories cross in the opposite direction. In looking at the type of recollective responses to an odor, younger subjects generate more "remember" responses, whereas older adults generate more "know" answers (Larsson, 1999).

Why does episodic-memory encoding decline with age? The answer is that the number of *new* episodes the individual encounters typically decreases over time. If an episode is a unit that contains a multitude of bits, and if it is determined principally by the sum character of those bits, then it matters not if the individual changes locations often, or even planets. Some bits will recur, making the resulting episodes more and more like preceding ones until there is no difference from one episode to the next.[22] Once a thing is "smelled," it is never smelled again, in the active sense. When the brain encounters an odor, it asks itself: Have I smelled this before? If yes: Was it good or bad? If it was good, then you are attracted; if it was not, then you are repelled. Conversely, if the brain categorizes the smell as "new," then it initiates the process of assigning memory, or encoding data, or "going live." By the time a person is 20 years old, the answer to the question of whether the brain has smelled the odor previously almost inevitably is yes.

When we join these two concepts – that semantic memories do not begin right away and only increase over time, and that episodic memories begin with the onset of memory proper and only decrease over time – it becomes evident that smell, the cousin of episodic memory, father or sister perhaps, is under less and less affront as a volatile, changeable memory complex. The point here is not to beguile the tenacity of smell, but only to put in order the relative probabilities in which the tenacity operates.

This discussion is becoming dense and thus warrants an expansion, or, rather, a recursion. Taking one point previously breezed over for further inspection: Youth, episodic memory, the limbic system, and emotions are a reinforcing multiplex. Speaking both phylogenetically and ontogenetically, prior to the use of cognition as a means of interfacing the world, the limbic system reigns. Hence, emotion, in tandem with episodic memory, guides the individual, decides for the individual. Smell is right in the middle of this process. The nipple, the baby's head, mutual satisfaction, limbic to the depths – the osmic sensorium is our first teacher as to what to avoid and what to advance upon. These lessons are valuable and thus are not easily modified or tossed aside. Episodic memory works like a pearl, growing in layers over time, encrypting the earliest memories with a string of code that gets longer with each successive layer. The first memories are the most

---

[22]*This can be related to the phenomena of ontological time dilation – the appearance of time moving faster as one gets older. When every new memory seems more like the last, there becomes a shrinking of difference, a relative loss of ability to discriminate between those differences, and thus a loss of attention to those differences. The perception of time is a matter of attention, where the less one holds, the faster time passes.*

densely embedded, and they project, prodigiously and increasingly, to new memories – semantic *and* episodic – as they accumulate, creating a web, dense, robust, and resilient.

Inviting its own dubious terminology to the occasion, smell is *limbodic*; that is, it runs on memory that has been coded into the body itself, via the limbic system. It commands its own network of physiological responses, taste aversion (via smell) being the most salient example. Memories "encoded into the body" are not as much plastic as they are cognitively created, semantically coded memories. Consequently, they are less subject to change. Abstractions and concepts, when they set the stage for the formation of memory associations, exchange and recombine. It is in their nature, for they are ephemeral.[23] *Limbodic* memory is embedded within the core living strategy modules that guide the individual in early life and in stressful situations throughout the lifespan (good and bad stress).

One last point to make here in support of the tenacity of smell-memory as a product of the limbic-emotional multiplex: The triumph of semantic memory does not finalize until the late 20s, when the prefrontal cortex is fully formed. This developmental stage can be understood as the inflection point between the two memory types. Coincidentally, the individual has then completed a seminal stage of formation, locking away a base set, a vineyard of autobiographical meaning, unconsciously created, pristine in its fidelity, and impossible to retrieve by will alone.

Fidelity and Lossless Data

We now enter an isomorphism of information storage and degradation; for the tenacity of olfactive memory can also be explored via this analog. A computer monitor is similar to the retina in that it can be thought of as a two-dimensional map, a grid of pixels, potentially embedding each cell of the grid with information. This is a reductionist view that for now is necessary to establish relative ideas; we will consider its limitations shortly. In the simplest way, when a black horizontal line cuts across a white screen, it is a string of "on" bits running through the map at a thickness of the single unit of squares on the imposed grid. To encode this image, each point (or square, as it appears) on the map is given an identity (A1, A2, A3 …). Each point is then assigned a value – on or off. That image, the black horizontal line, exists as a state of the map in which certain bits are on and others (most of them, in this case) are off. This state embodies an amount of information in proportion to the length of the line (and technically, the size of the entire bitmap within which it exists).

---

[23]*Technically,* color *is a concept (and so is* tone*). It only exists relative to its position within an established set, and that set, or spectrum, or scale, is in itself a concept that must exist as a substrate for the meaning of the stimulus to be identified. Color, and vision in general, is closer to being a cognitive, conceptual phenomenon compared to smell – again due to its phylogenetic developmental positioning. And thus color memory is less present in the episodic, limbic realm.*

This is the basic idea of pixels, the picture elements being the units, or the bits of the map that are either on or off. The retina can be understood as a similar two-dimensional map where the bits or squares or points are photoreceptors, and they are either activated by a photon or not. The bitmap as a form of storage makes use of its structural correlates to the retina and its subsequent neural processing. This is nowhere near the level of complexity required to describe the phenomenon of vision. For now, however, it will suffice. And to its limitations we now turn.

Vacillating between analogies, assume the brain has a limited capacity for storage. Saving an entire 100x100-unit bitmap for a line that is 1 unit thick and 100 units wide seems like a waste of storage space. Instead, the brain employs "compression," a way of simplifying the massive bitmap into a simpler, smaller Line. Some kinds of compression, however, are subject to data loss and are thus called "lossy." Because of the desire for information conservation in regards to storage and the use of lossy compression, every time a picture file is opened (decompressed), edited, and re-stored (compressed), it is further distorted. Given the vast number of bits that are either turned on or off to create the line, some of them are bound to switch, in what can be called generational degradation. When there are too many bits, it is unrealistic to expect that they remain in perfect fidelity upon repeated retrieval, use, and storage.

Consider the memory as functioning in a similar way. The semantic bits, the stimulus bits, the contextual or episodic bits; they are all encoded into the memory file. It has been previously postulated that the episodic bits are unique in that they are less plastic. This concept makes the analogy confusing, because it pushes the associative network paradigm in front of our monocle. Instead, this example shows memories of individual, isolated units being retrieved and re-stored in the process of perception. These memories are bound to distort over time, losing definition (though acquiring a new one, in the course of memory), unless they are never retrieved in the first place, as is the case with unconsciously coded odor stimuli. A smell-memory encodes multiple bits of information into a single access point, but it provides no means for intentional retrieval.

Olfactive memory is not encoded consciously. It precedes (subjective) consciousness, both in phylogeny and ontogeny. Moreover, because it is encoded unconsciously, it cannot be retrieved with the inverse; that is, it cannot be summoned by will alone. It *can be* activated by a stimulus, but not by intentional self-solicitation. Smell is nothing like an image file, but if it were, it would be one that could not be clicked on. Instead, it would open by itself, and only under very specific conditions. And so, when a memory is never conjured intentionally but is activated only by a 1-in-a-10,000 chemical, or, rather, a one-in-a-million combination of said chemicals, then tenacity seems more like tranquility. Our smell memories sit like a flower, with all of its pollen still stuck to its stamen. The bee rarely comes, and the pollen clumps upon itself, waiting to explode on contact, to hijack your body and intoxicate your mind with indiscriminable simulations.

4. Weakness

Humans differ from all other creatures in that they interact with their environment primarily via a cognitive apparatus. This reality signifies a shift beyond the physical world into the metalogic realm of meaning making. Perception is the process whereby the mind makes meaning of sensory information. It can be thought of as having two modes of operation. The first mode functions on the biological level, where the meaning of the perceived stimulus is translated directly to the body. The physical stimulus generates a physical response within the organism. In the second mode, a meta-meaning is created: The biological meaning *is* the stimulus, and cognition is the response. Hence, this mode makes meaning *of meaning.* These two modes of operation can be referred to as biosemiotics and cognitive semiotics, respectively.[24]

On the level of biosemiotics, pheromones are perceived as powerful messages that elicit a cascade of direct sensorimotor responses, as well as the gamut of corresponding physiological responses. The buck in rut does not cognize the scent of the doe. It responds. There is no deliberation on the course of action, only a phylogenetic algorithmic circuit of yes-no's where the switches are not clicked by "the deer" but, instead, by its deer-ness. Cognition, by contrast, allows humans to detach the stimulus from its biological imperatives and reassign new meaning for the body to act upon. The immediacy of pheromone stimuli, for example, is put on pause while the cognitive apparatus assesses the stimuli using its own criteria.

Given the supremacy of cognition, predetermined biological responses to stimuli can be overridden by the higher-order meaning constructed with thought. Humans interact not only with things but also with their ideas of things, and especially with the ideas of others. Olfaction is, primarily, a biosemiotic system; that is, it elicits a physiological response to chemical stimuli. Cognition intercepts this response, extracting the stimulus from its physical interchange and modifying its "meaning" according to the nonphysical, simulated representation of the mental world. Via this cognitive interception, the cues and suggestions that are attached to an odor stimulus can have more impact on perception than the chemical properties of the stimulus itself. Cognition overrides biosemiosis.

Again, the concept of evolutionary time comes into play. Olfaction developed before cognition; its naturally selecting design objective was *not*

---

[24]*A note on "cognition": Herein, this term refers to or names those general cognitive activities typically associated with things like language, semanticity, and overt, explicit analysis; in a sense, it is more about the controllable parts of cognition (and their subconscious correlative operations), the parts more typically thought of as the distinctly human.*

to collaborate with cognition for meaning making but, rather, to generate meaning by itself. (In a way, olfaction is a kind of primitive cognition, but one that is surely redirected by the later advancements in the reciprocal interactions of organisms and their environment.) Because olfaction was built on a noncognitive platform, it is subordinated. The strengths of its messages are outweighed by the cognitive imperative placed on human interaction, because it did not have time to develop a resistance to this cognitive interception.[25]

Hackability

In a slightly incorrect usage of an already ambiguous term, cognition *hacks* the olfactory system. This is not to say that smell is not capable of having a strong influence on human behavior. Rather, it suggests that its influence pales in comparison to that of cognition. In fact, even those senses that have a stronger link to cognitive functions yield prioritized information that can seemingly manipulate our perception of smell. Often, smell must corroborate with other senses for validation. To cite one prominent example, visual semantic clues can radically alter an individual's perception of an odor.[26] For example, it has been reported that coloring white wine red results in the wine taster describing the wine with words more often used in the description of red wine (Lawless, 1984). In other words, if the wine *looks* red, then it *tastes* (and *smells*) red.

Enter the most simple and effective experiment for testing the fallibility of our olfactory apparatus. All you need is some cotton swabs, distilled water, and a general audience of any size.[27] Inform the audience that you are trying to measure the speed with which a smell radiates from its source and that you will now present them with a smell that will be unfamiliar to them, but noticeable. Dip the cotton swab in the water, and hold it up. Ask the audience to raise their hands as they smell it. Then, do not simply watch as the audience raises their hands at "smelling" the odorless distilled water. Instead, ask them to write it down and then later read to your own amusement the list of completely unrelated odor apparitions. Olfaction is simply not an explicit activity. Köster (2002) argues that fragrance and

---

[25] *This all seems to stand in clear contradiction to the earlier section on the limbic system, where the olfactive-limbic relationship prevails, temporally at least, over all other sensory perception processes. Then it must be noted: because humans live in a cognized world, stress itself is internalized in the mind, and is to be dealt with virtually. The limbic system is only in charge during stress-induced situations. For all else, problems are virtualized, and thus dealt with in a more "efficient" manner.*

[26] *Semantics is the identifying, labeling part of language, and is often juxtaposed against syntax, which is the organization or arrangement of those words – vocabulary vs. grammar.*

[27] *Supposedly, this experiment was first performed in 1899 by Emory Edmund Slosson, professor of Chemistry at the University of Wyoming (Hertz, 2005).*

flavor research panels are incorrectly queried about the product directly when they instead should be mined for their own observable behavior. Asking the participants to think about the product is to, unnaturally, make the implicit explicit.

A bit more difficult to prepare, this next experiment was at least well documented. In 1978, an entire listening audience was induced into outright olfactory hallucination. Michael O'Mahoney, now a food science professor at UC Davis, announced via live radio broadcast that particular frequencies on the electromagnetic spectrum can initiate olfactory perception.[28] The radio squeaked out a tone, the odorous apparition was released, and the station was thereafter delivered listener mail recounting their experiences with the induced olfaction. The audience reported a broad and randomly associated list of odors, with some people even developing migraines as a result of their aromatic apparition (Hertz, 2007). Granted, "As in vision and other sensory systems, expectation and internal behavioral state can influence odor perception" (Wilson & Stevenson, 2006: 34).

It is also known that unfamiliar odors, and odors for which the source is unknown and therefore beyond control, are by default perceived as unpleasant (Engen, 1982). This phenomenon works in concert with our poor use of language to identify smell. Because we are language-savvy creatures, verbal labels have become the dominant mode of identification and thus of separating the known from the unknown. Without these labels, and the identification purpose they serve, we are more likely to be led into such unfamiliar and potentially frightening situations. The fact that smell is such a "hidden" phenomenon (Köster, 2002), and is never given an explicit opportunity to be consensually or interpersonally recognized, makes it then even less likely that people would talk to *one another* about their experiences, to cancel out the possibility of their overreacting, for example.

Inversely, and yet amplifying the radio hallucination phenomenon, it appears that people (unconsciously) judge the healthfulness or harmfulness of their environment based on smell (Cain, 1987). This occurs despite the fact that toxicity does *not* necessarily carry an olfactory signal.

Something is being hinted at here that will be expounded upon later, but that for now must be stated clearly. Smell is a learned perception: It is not hardwired into every human in the same way, eliciting the same responses. It is our subjective life experience that informs our response to odor, and not the collective experience of our species. Pheromone olfaction acts in this way, but not "normal" human olfaction. Because we must *learn* how to respond to smells, because cognition is such an integral part of how we

---

[28] *This experiment was revisited by Knasko et al. 1990, who studied the effects of feigned odors on subjects' well-being etc.*

learn, and because other senses inform cognition in a more direct way than olfaction, smell is a very easily hackable sense.

Although it could be competently argued that animals learn just as humans do, for purposes of discussion herein it must be recognized that humans exercise an entirely different form of learning, so different, in fact, that all nonhuman learning might start to look like something else in comparison. Human learning – henceforth simply "learning" – is a visual-linguistic phenomenon.

Vision is the most recent sense to fully develop.[29] Even later, the development of language as a kind of cognitive sense, a metasense, comes to us in the form of visual, auditory, and even tactile stimuli (in the case of Braille reading, for example). Therefore, these two modes of perception – vision and language – have been adapted over time to work both with each other and with the cognitive apparatus responsible for our learning. Our brain thus prioritizes visual and semantic information, leaving anything that comes to us via another channel subject to manipulation.

In other words, what a thing *looks like* and what it is *called* are more important indicators of what the thing *is* – even if that thing is an odor. What an odor *smells like* is actually less important when it comes to our perception of smell.[30] The weakness of smell stems from the single stalk of our very existential ontology – we are highly cognitive creatures. We have created an entirely ethereal world of thought in which the molecules of our universe simulate their interactions. We live in a world above, with smell below, an artifact of our ancient geotropic origin.

Fatigue

Any exploration of the weaknesses of smell would be remiss in neglecting odor adaptation. Under Engen's experimental conditions, published in 1991, the olfactory system is neophobic, and it exhibits a high false alarm rate. All new odors evoke "apprehensive arousal," and the longer they remain unidentified, the more defensive the behavior. The sense of smell is a "vigilant detection system" following one rule: When in doubt, it assumes there is an odor and the odor is strong (vs. weak). As Engen, via

---

[29] *To be specific, "human vision" consists of 1.) binocular vision allowing spatial depth perception, 2.) tricolor vision receptive to the full spectrum of visible light, and 3.) the expanded field of vision resulting from the raised vantage point of bipedalism.*

[30] *On this point however, it should be noted that many studies of odor identification can be criticized upon the fact that the subject's response uses an auditory channel, which is habituated to cognitive direction and semantic override, and thus might be changing only the subject's reporting of perception, but not the perception itself.*

Green & Swets (1974), explains, the cost of missing an odor signal is more serious than raising a false alarm. This process reverses dramatically upon repeated exposure to an odor, after the odor is no longer "new," and once the "false alarm" has balanced out its asymmetric payoff function.

Odors stimulate only our memory; they do not cause any other reactions by themselves. When we feel sick in the presence of an odor, it is *not* because of the odor itself but because we associate it with previous sickness. Automobile exhaust smells and is deadly, but it is not deadly because it smells – carbon monoxide, the "silent killer," is odorless. The odor sensorium has no capacity for discriminating toxic from nontoxic substances. Odors are epiphenomenal, without any direct biological effect other than its effect on the olfactory system. Still, the powerful illusion that odors cause pleasure and pain persists. Even when a person knows that it was drinking too much whiskey that made him sick, the smell of the whiskey remains aversive as though it were the cause. Because smell-response is learned, it makes intuitive sense that upon repeated exposure to a smell, noxious or not,[31] its alarming effect will wear off over time. Why else would you still be in its presence – "It mustn't be dangerous." The effect is thus turned off, like a smoke alarm that you disable when you accidentally burn your breakfast. Cognition is practically *defined by* its ability to move beyond this reliance on false alarms. It generates methods of identification completely unattainable in the detail and at the distance (both spatial and temporal) that olfaction offers.

A great anecdote to reveal this odor fatigue is the story of Febreze.[32] Febreze is an odor-eating spray intended to deodorize a smelly room. The product did not work as planned. As it turned out, its failure resulted from this salient behavior of smell – desensitization, attenuation, and adaptation.[33]

Febreze (i.e., the mathematicians, habit specialists, and greater marketing team of Proctor & Gamble) even tried to teach their customers how to use the product via television commercials: cue (harsh smell) – routine (spray) – reward (smell is gone). However, it didn't catch on. The product was a

---

[31] *This is not to be confused with a "smell" that is trigeminally noxious. Mustard gas stimulates trigeminal nerves, not olfactory nerves. In fact, a good way to disambiguate smell from trigeminality is to determine whether it physically hurts. Smells do not hurt, they* remind *us of pain, but they do not physically* cause *pain.*

[32] *This story is taken from a news article about predictive analytics, see Duhigg, 2012.*

[33] *"Attenuation" presumes a medium, and is more commonly referred to in electromagnetic phenomena, not chemical. A loss in olfactory intensity does not arise from its propagation through the medium* per se, *but from its spatiotemporal expansion within it. Even the concept of intensity loss in olfaction is vague, considering the substantial variation in perception thresholds for different chemicals/odors. The "attenuation" in use here would be more applicable to the information sciences, where the signal is distorted due to noise in the channel. In the case of olfaction, the signal (odor) is repeatedly reassigned to its false alarm value (due to the fact that the subject has not left its proximity), and each false alarm lessens the intensity of each subsequent signal, thus "sharpening the perceptual field" (Gottfried, 2009).*

failure. The team then proceeded to conduct field research. In the course of this research they encountered a neat-freak of a woman, living in a home suffocating with the smell of cat urine. *She* was not gagging. Just the opposite: She had adapted, and she could not smell a thing. The team eventually discovered dozens of other homes like this one. Their product did not work because their customers were missing the initial cue – the bad smells. Febreze eventually learned to rearrange its "instructions" so that the cue is now the *visibly* messy room, the routine is *cleaning* the room, and the reward is the satisfying aroma of Febreze (a product originally designed to eradicate smells).

Our sense of smell shuts off quickly, just as staring at a bright light disables our vision. Any persistent odor that does not instigate movement of the sensing organism is reduced to noise in the channel. The smells of our homes, our bodies, our friends – they all become invisible to us. Furthermore, when we are asked to focus our noses, the scenting apparatus defers to the higher senses for verification. The olfactory system thus has its weaknesses, and it is preyed upon indefensibly in the domain of cognition.

Part Two – CONTRADICTION

The psychology of olfaction is experimented on using at its base only semantic memory. This process creates many problems, the most imminent of which is that in order for subjects to participate in a study, they must *verbally identify* the odor. However, the language of smell is very fuzzy: It is not explicitly taught, and it is not made identifiable by any measure of consensus. There is an elephant in the laboratory. Also, the way we learn about odors and the way we learn language are two very different processes that rarely cross paths in ontogeny. To date, there is no comprehensive system for classifying smells.

Two major entities maintain the crypt of olfactory description. These gatekeepers of odor information are the fragrance industry and the scientific discipline of organic chemistry. Each has its way of making it very difficult to make sense of the smells around us – and in particular the aroma compounds that make them up.

Chemistry is the logical progression of alchemy, which is more like a hybrid of religion and science. In this way, the history and the vocabulary that comprise organic chemistry are filled with mythologies and half-truths. Nostradamus is credited with developing the first method for making benzoic acid, the precursor to the benzene ring, a concept that characterizes the entirety of organic molecules. Further, the concept allegedly was revealed to chemist August Kekulé in a dream of an Uroboros composed of a chain of carbon atoms eating their own tail.

Of all the chemicals there are to know about, the only ones we can smell also happen to be the most complex. The subdiscipline of organic chemistry was not initiated until well after the establishment of chemistry proper. The study of organic molecules therein falls so far out of the range of inorganic chemistry and yet is still far enough away from biology that it is in need of its own category.

Somewhat like their historical origin, aroma compounds are volatile. They can change into other compounds under the right conditions, some as simple as oxidation. When these compounds are isolated and stabilized, they can be identified by their properties and hence can be given a name. But this is where the problems really begin. Of the infinite number of organic molecules, each one can have almost a dozen names.

Let us view one as an example: sotolon, the smell of burnt sugar, maple syrup, curry or fenugreek, and a component of coffee aroma, roasted tobacco, and celery. It is officially known as *4,5-dimethyl-3-hydroxy-2,5-dihydrofuran-2-one*, but it is more informally called caramel furanone, sugar lactone, or fenugreek lactone. It has a "formula name" of $C_6H_8O_3$, which identifies its molecular components. There are even "names" that take the form of 3-D structures, required by the massive complexity of these molecules. We can only imagine the margin of error in some publicly available repository of odorant information. If it is clarification that we seek through the lens of chemistry, then perhaps we should reconsider.

The other gatekeeper is different. The handful of companies that make up the fragrance industry operate with great power and airtight secrecy. The discovery of new aroma compounds, and of cheaper methods of replicating known ones, is a critical endeavor for their business. However, keeping these discoveries unknown to competitors is even more critical, because there are no intellectual property protections for fragrance or for chemicals. In instances where a company has synthesized a significant aroma, it provides a trade name – but that is all. It does not release any other chemical clues so as not to inform its competitors. In fact, in the case of the synthetic musk Phantolide, the international body of chemical nomenclature had no name for the molecule for years after it was discovered by the fragrance industry. Anecdotally, many of these "discoveries" by the fragrance industry were actually accidents of other pursuits, such as explosives (synthetic musk) and antidepressants (Calone – "the smell of the ocean").

Fragrance companies do not supply the public with information about aroma compounds. Instead, they advertise fragrance – artfully crafted combinations of aroma compounds. And, in the case of masking fragrances, where the intent is to use a good smell to camouflage a bad smell, the scent is invisible by design. Consequently, not only is it in direct opposition to the business model of a fragrance company to divest information about its products' ingredients, but their desire and capacity for discovering new compounds are greater than those of publicly funded, publicly accessible scientific discoveries.

Outside these gates, in the pool of public knowledge, smell is a symbol for confusion and fantasy. It remains among the common people in a way that has not changed for hundreds of years. Smell is, after all, the animal inside us, so it spends very little time in collective discourse. Smell, in the everyday sense, does not lend itself to contemplation; in fact, it tends to avoid language altogether. People may talk about what smells *mean*, or where they *came from*, but they do not talk about the smell itself. People do not converse about the constituents of an odor. In the vast majority of cases, a smell is referred to as a source, not as a molecular profile of potentially interchangeable parts. Beyond that, people do not consciously notice most smells in the first place. Olfaction operates automatically, actively de-noising and investigating every one of the myriad volatile organic compounds of our vaporsphere so we can do more important things, like conducting chemistry experiments or creating new fragrances.

Furthermore, olfactory perception is so riddled with byzantine circuits that it leaves itself highly susceptible to cognitive override and subjective distortion. By the time a smell finds its way to a person's conscious awareness, it is far removed from its "universal" essence. It is somewhat of an overstatement, but not by much, to assert that the general public barely knows what smell is or how it works. Any discussion on the topic shifts immediately to its effects on memory and emotions, leaving the objective analysis of odorants a fallow field.

The etymological history of smell is shrouded in confusion: The public does not know enough about smells in the first place, and the authorities on odorants make it very difficult for the public to *correct* their confusion. Following this explanation of the language of smell, the stage is set, and we are now faced with the difficult task of giving order to our olfactory experiences.

The language of smell is to semiotics what the Uncanny Valley is to our empathic mirror neurons, a kind of lexicographic epilepsy. Although the nature of odor classification *implies* semanticity and its logical correlates, it is informed by the limbic language of olfaction. Its study requires a circuitous path, one that is already quite complicated. The contradictions inherent in olfaction as an information system are many, and only through the repeated arrangement and rearrangement of its approximate parts can we begin to organize the cloud of olfactory data.

5. The Language of Smell

First and foremost, smell is about the furthest thing from a language that we can get. The two have nothing in common, at least not according to the way they would tend to be described. The two make an odd pair. Smell is right-brain dominant. It is so tied down by its primitive system of agency that it does not have a chance to integrate into any existing lexicon. (The language centers of the brain, the Broca and Wernicke areas, reside in the left hemisphere.)[34] The seat of olfaction, the limbic system, has nothing to do with language. Screaming in horror is limbic – but that is not language (Sapolsky, 2010).

Furthermore, smells are not explicitly learned. Being that most verbal exchanges regarding smells do not progress beyond the good/bad hedonic dyad, we do not take the time, nor make the mental effort, to analyze what we smell for its constituent parts, its similarity to other smells, its super- or subordinate categories. Because of this absence from social discourse, there is less likely to be a solid agreement on smell names. And, because of olfaction's primitive processing, it is even less likely to be used with descriptive language. Instead, it favors guttural utterances and associated adjectives. With no further delay, and utilizing the only "language" available, to the *Lingua Anosmia* we turn.

Surveying the Linguistic Territory

Hopefully it has been already established that smell makes a mess of language. In order to shift the discussion from olfaction to language, a few introductory remarks must be articulated.

Disambiguating Names and Things

Let us pause for a moment and revisit the basic question: What is "Smell?" It is a process, a reaction, a verb. It is a phenomenon, an experience. And, it is, of course, an odor molecule, an aroma compound; it is evaporated, a vaporous substance wafting through the air. Evaporation requires heat. There is garbage, and there is hot garbage. A chemically mediated phenomenon, smell is inescapably associated with its source. It is because of this link that smell suffers from the name-thing problem. "Orange" is a

[34] *This is not to say that cross-hemispheric communication does not exist, because that would be the furthest thing from the truth, only that neural anatomy does play a part in the integrative abilities of different modules.*

rich example of this problem, because it establishes a name-thing relationship amongst multiple nodes of the lexicography.

Oranges, and all citrus fruits in general, were unknown to the ancient Greco-Roman cultures, because these cultures were isolated from China and its citrus-rich environs for the duration of their existence. Consequently, the word *orange* did not enter the European languages (via Old French) until circa 1300 C.E., and the color did not appear until hundreds of years later.

The word "orange" is polysemous: It means a few different things. It is the name of a color, one of the gradations on the spectrum of visible light. It is the name of a place in France from whence the Anglicized color-name is often misattributed. Orange is also the name of a fruit, derived from Romance-speaking traders who called it something similar. Finally, it is the name of a particular scent, one that comes from the aforementioned fruit tree. As with many smell names, the "smell of orange" is usually referred to as the orange fruit itself, which is made up of many molecules, some of which are perceived as odors and have their own names, such as limonene, myrcene, and alpha-pinene. (Note: Orange-the-smell is not one particular molecule, but a combination of molecules.) Technically, the problem is that in referring to orange-the-smell, one is also referring to orange-the-fruit.[35] Gardenia is a flower and a smell, and so are jasmine and rose. Further, some of the odor molecules that make up an orange also make up a gardenia, and so on.

And yet, orange might be referred to as lemon, in that the two share the superordinate category of citrus. From a subjective perspective, and because human test subjects perform so poorly on smell-recognition tasks, there tends to be an indiscriminable difference among many of the citrus smells, especially in regards to this orange-lemon relationship (Dubois, 2007).

Herein, reference might be made to the orange fruit itself, or to the orange smell emanating from that fruit, or to the primary constituent molecule that distinguishes that scent.[36] However, this distinction may not be overtly articulated. For this reason, the context of the word use is often critical. Smell offers such name-thing conflations, and disambiguation amongst them all is cumbersome and makes for difficult reading, especially upon the subject of Language. It is therefore rejected in favor of fluency.

---

[35]*In total, the orange tree avails three different smell-names: the "Neroli" flower, the peel of the orange fruit, and the leaves, called Pettigrain.*

[36]*Although limonene shows the highest concentration, as per gas chromatography analysis, this does not indicate the subjective perception of individual molecules according to their prothetic measurements, that is, their intensity. Though some other molecules besides limonene occur at lower concentrations, their role in creating the scent's orange-ness may be equally as important. Smell is a holistic perceptual phenomenon, after all.*

Speaking in Tongues

This discussion will be generally limited to the modern English language. This is a very loose boundary, because many smell names retain their native word form. Ylang-Ylang, a flower of the Philippines, is called by no other name, adding Tagalog to the interlingual lexicon.

The most important tongue in the world of smell is French. Much of the language used for smells either originated from or was bolstered by the perfume trade, and, by extension, the sources of those aromas, and, once further, the whole macrocosm of global trade circa 1700. It was a time of an awakening social order, a "coming to civility," and France was at the forefront.[37] It is of this time in history that Alain Corbin writes his book, *The Foul and the Fragrant*, that considers the role of odor in French society. There is no coincidence, although it is rarely considered, that "to be civilized" meant to *not stink*. With perfume, one could at least pretend. And so, through an act of historical congruence, many words of smell in English, as far as they can be related to perfumery, come from French.

This is not to neglect, as mentioned, the *sources* of those materials. These could have come from places the world over. The Far East and the Arabian Peninsula, for example, generate much of the sources of the aromasphere and thus have some part as well in generating its names.

The Nature of Etymology

Language, like history, is full of half-truths, mixtures of what is, what was, and who's speaking. The names of smells have followed the same mindless route, across the globe, throughout history and by the power and guidance of a collectively anthropogenic force with no intelligent foresight or intention of order.

In the development of a word, during this process called Language, there is a period where the word is molten: It behaves like a liquid metal. It gets twisted and stretched, deformed and re-pressed – but only as long as it stays hot. Eventually the metal cools, and no matter what its final shape, there will be no more changing.[38] In his history of fragrance, Edwin T. Morris (1984) recounts such a tale involving benzene. The word was mined from the cultures of the Far East, the Middle East, and North Africa. The Arabs called it *luban jawi*, meaning "incense of Java." Via India, the word has simply become *loban*, meaning "incense." It then made a categorical leap, using the superordinate to refer to its sub. This *luban jawi*, in fact, is capable of making such a leap, because it is such a good olfactory representative of the superordinate, incense.

---

[37] *Red meat is called Beef (or Boeuf) by the French aristocracy being served, and Cow by the English peasants who are still in direct contact with the animal.*

[38] *Or at least, not without much "effort."*

The European traders heard this *luban jawi* as "ban-jaw-i," twisted it while it was still hot, and made it "ben-zo-in." Of the thousands of European traders who interacted with their Arab counterparts, in hundreds of thousands of transactions, the word flipped and flopped, back and forth, cooling off, one transaction at a time, until the most cross-cancelled version is revealed – cold, hard, and making no sense at all in regards to what the thing is.[39]

Furthermore, the "gum of the benzoin tree" is the common name for a natural substance, which occurs as a resin in many species of tree in the *Styrax* genus (not limited to one species). Benzoin-the-chemical does not, according to etymological logic, come from benzoin-the-resin. Instead, it comes from benzaldehyde – a chemical derived from bitter almonds.

What "gum benzoin" does yield is benzoic acid, which was called "flowers of benzoin" upon its initial intentional crystallization, generally credited to the distillation technique performed by Nostradamus in the 16th century. This, of course, is a superordinate exchange, in which the category of Floral, intimated by the term "flowers" of benzoin, is exchanged for that of its native smell category, Resinous. From a perfume-based mode of classification, these two categories rarely overlap. This observation suggests placing the metaphorical genesis of the term upon the image of a crystal that blossoms like a flower upon distillation to reveal its scent, and less like the odor-specific, metathetic qualities of the benzoin aroma itself.

Finally, we arrive at benzene, a chemical compound derived from benzoic acid, although not originally from a *Styrax* resin. Benzene was derived from coal-tar, a substance familiar to organic chemists of the 19th century for yielding many organic compounds. The most simple, symmetrical, and stable of the double-carbon bonds, benzene represents an entire chemical family of similar molecular structures made of closed rings of bonded hydrogen and carbon atoms. This family, again represented by the ideal form of the benzene molecule, is referred to as the aromatics, because, like their prototype, they tend to be pleasant-smelling.[40] In chemical nomenclature *–ene* is taken from *ethylene*, the simplest of alkenes, another chemical family, which is related to the aromatics in that they are characterized by the nature of their covalent carbon bonds. Benzene fits this nomenclature requirement; hence the suffix *–ene*, and the final quenching of the once-molten metal.[41]

---

[39] *Similar, and yet not, to the birth of the word "ampersand," the name of the ampersand symbol (&) is said to have come from a convolution of the words "and per se 'and'," uttered at the end of the recitation of the English alphabet, at a time when '&' was still the 27th letter. While it was still 'molten-hot,' the name of the symbol technically changed from "the 'and' symbol" to "ampersand." It is not similar to Benzoin in that it does relate linguistically to its origin, fragmented as it is, as a conflation of the phrase "and per se 'and'."*

[40] *More on the subject of "aromaticity" below.*

[41] *Benzene may also be called benzyne, or benzoic acid, depending on the permutation, how it has been synthesized, or how volatile it is in its respective environment. Also note; Benzene is*

Take a moment to retrace the path, back from the aromatic benzene to *luban jawi*, for this is the characteristic trajectory of the names that have been "given" to the actors in the osmic phenomenopera.

The Chemist's Language: Organic Chemistry Bites Its Tongue

At the outset, fragrance writer Edwin Morris reminds us that the olfactory sense has always been an important tool to the chemist (1984). It is no wonder. Smell is the "chemical sense," and, to date, it may be the most effective measurement tool we have for odor detection (in vast opposition to vision, for example). Hear the lamentations of this author, an olfactory scientist of the 1960s: "Time has its seconds; temperature its degrees (Kelvin); loudness has sones or bels; brightness in brils, but no "olfs" yet can bolster the science of smell. ... I want our work to be scientific, too" (ASTM, 1968: 17-18). Smell is a sparsely populated landscape in regards to the kinds of measuring tools scientists are familiar with.

Although the *dynamics* of odor chemicals with the organism lack their share of measuring rods, the odors themselves are an integral part of knowledge acquisition in the field. Describing olfactory phenomena during chemical experiments and naming the source of those odor perceptions is critical to chemistry. Significantly, major achievements in this area have been made over the past two centuries.

If the language of the fragrance industry is French, then the Lingua Franca of the chemistry of smell is ultimately Latin, the natural choice for all scientific taxonomies. Following the revolution in organic chemistry that occurred roughly during the 1850s, this universal language-base for olfactory phenomena shifted from the botanical to the chemical nomenclature of the International Union of Pure and Applied Chemistry. Though this IUPAC system is Latin-influenced, most of the actual word construction is more a function of the nomenclature rules than of Latin etymologies. Fluency in both of these languages is a prerequisite for a neat disentanglement of the instigators of the olfactory world. However, fluency, in that sense, is not the purpose of this exposition.

The history of fragrance, and of chemistry in general, is twisted by the practice of alchemy, a pre-science, a just-so but not really science where recipes and results alike were based in part on ephemeral constructs of metaphysical constants and vernacular lexicon. Clearly this is not the most

---

an important component of gasoline, and is substituted for the name gasoline in many countries.

conducive crucible for a linguistically sound classification system. Nonetheless, chemistry has a language to itself, as well as a history that reads like an 18th century science fiction novel (if they had existed at the time). One would expect chemistry to be clean, precise. In reality, smell is *organic* chemistry, and any of the traps of chemical nomenclature are multiplied when they are placed in the organic arena. It turns out that what can be called the birth of organic chemistry itself was one of these fantastic and fictive-sounding events. This is the discovery of the aromatic ring structure of benzene, which is retold below.

Despite his many years of working as an organic chemist, German scientist August Kekulé nevertheless denied his agency in the discovery.[42] It is reported thus: Kekulé was perplexed by the molecular structure of benzene, as were many chemists at the time (circa 1850). The empirical formula of the molecule was known; it is made of six carbon and hydrogen atoms. How these atoms are adjoined was the puzzle. How could this substance contain so much carbon and yet be so stable? Carbon is tetravalent, meaning it needs to form four bonds to be content. Kekulé referred to these bond requirements as "affinity units;" they are now known as "valences." If each carbon atom in the benzene molecule is bonded to only one hydrogen atom and at most one other carbon atom, then what about the other bonds? They are left empty, and the remaining carbon atoms should be extremely eager to latch onto any free electrons they can find. Kekulé could not wrap his head around this dilemma.

One afternoon, while reclining by the fire, Kekulé lapsed, and he envisioned a peculiar commotion of carbon atoms, dancing. Rhythmically, the structures, the undulating knuckulars, grabbed their own tails, to stick in their mouths – the Uroboros is initiated (see Figure 5).[43] Upon coming to wits, it was revealed to Kekulé that the perplexing stability of the grouped carbon molecules could be due to a closed ring formation, each carbon atom interlocking with the other, in a hexagon of six atoms. More specifically, each atom alternately bonds and double-bonds with its neighbor, and also with the corresponding hydrogen atoms along the perimeter. This structure fulfills the four-way stability requirement of the carbon atoms. This "benzene ring," as it would later be called, was easily identified by its scent, unmistakably reminiscent of the gum of the benzoin tree.[44]

---

[42] *For added narrative appeal, it should be noted that this discovery was made in simultaneity with another scientist, (although this is a common occurrence in scientific discovery).*
[43] *The Uroboros, the snake swallowing its own tail, is the most dominant symbol of paradox and of (self-reflective) consciousness itself, and perhaps the most archetypal of symbols to arise in human culture.*
[44] *Although it was not, in practice at this time, derived from the tree, but, instead synthesized from coal-tar in a laboratory. Etymologically, only by way of its similar smell, is Benzene related to Benzoin resin.*

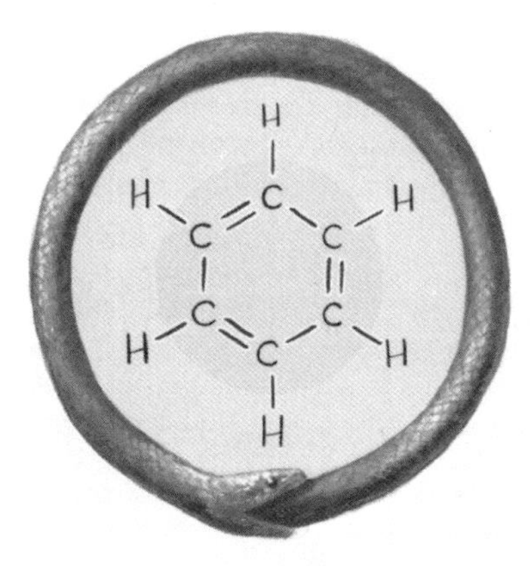

Figure 5: The benzene Uroboros represents the double bond of carbon atoms that make up the Benzene molecule. These atoms bond to themselves just as the snake swallows its own tail. Prior to the discovery of this structure, chemists could not explain how the molecule remained so stable despite the tetravalent requirements of its carbon atoms. *Illustration by Joe Scordo.*

When Kekulé announced his discovery of the self-linking benzene ring, it was collectively transmutated into the *aromatic* benzene ring. It is here that Kekulé's aromatic structure bifurcates the field of chemistry, and it is here that the first point of classificatory confusion rests.

Smells, or volatile molecules, are informally called *aroma compounds*. This makes sense, because they are (obviously) aromatic, and they are certainly compounds. Individual atoms can be called chemicals, but they should be more specifically referred to as chemical *elements* to differentiate them from chemical *compounds*. When more than one atom-chemical comes together, the resultant molecule is called a compound. Water is a compound, as are all smells. Sulfur atoms, for example, are odorless until they combine with hydrogen to become hydrogen sulfide, the smell of flatulence.

The fragrance industry uses the term *aroma compound* to refer to any odor molecule. By contrast, chemical science uses it more specifically to refer only to those molecules that resemble the ring structure of Kekulé's benzene. This is an initial example of the confusing taxonomy constructed from this, the Uroboric animal that is the chemical language.

The general nomenclature of chemistry is a system of naming, developed over hundreds of years and by a multitude of chemists, all speaking a variety of their own natural languages.[45,46] This language of chemistry,

[45] *The body responsible for this universal nomenclature is the International Union of Pure and Applied Chemistry (IUPAC), formed in 1919 (IUPAC, 2013). Note: The need for such an organization was voiced by none other than August Kekulé, around 1860, the complexity of organic chemistry being the main impetus for its creation.*

simplified to the chemist, is highly complicated and all but unreadable to the layperson. Any particular odor molecule could have up to a dozen "synonyms," with varying levels of veridicality depending on a number of chemically sensitive factors. Organic chemistry is so complex, in fact, that a visual language of three-dimensional structures is required in order to study it thoroughly. (The most remarkable of these structures is the DNA double-helix.) This language works *within* the field of chemistry. Upon export, however, its linguistic products become, at times, a point of ambiguity rather than one of explication.[47]

Let us take isomers as an example. A helpful way of identifying and categorizing these complex, macromolecular organic compounds is by their structure – the types of atoms they are composed of, the shape of their arrangement, and the relationship of the bonds between them. Many groups of aroma compounds have similar structures or functions, or both. However, there are cases where seemingly similar structures produce different smells. Regardless, these compounds are called *isomers* (see Figure 6). The name conveys the fact that the components are the same – that is, they are composed of the same type and amount of individual molecules – but they are arranged differently. Different isomers can have very different smells, thus making somewhat of a mockery of the nomenclature when it is used to identify odor properties. Each isomer requires an extra prefix to be added that determines its specific instantiation. The compound carvone, for example, is formed as either the R-(–)-carvone, which smells like spearmint, or the S-(+)-carvone, which smells like the caraway seed.

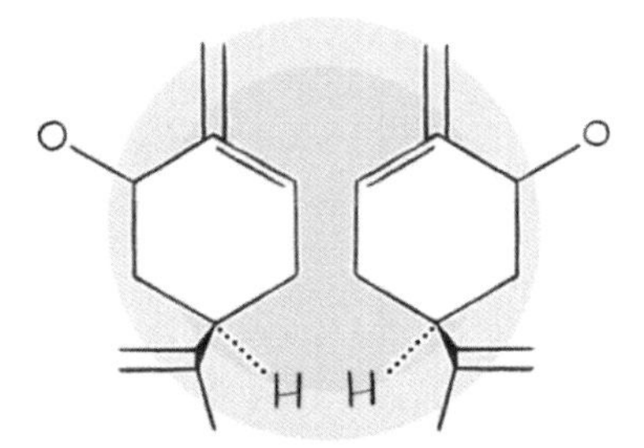

Figure 6: The isomer carvone exists in two mirror image forms. R-(–)-carvone smells like spearmint. Its mirror image, S-(+)-carvone, smells like caraway (Leitereg et al., 1971). *Illustration by Joe Scordo.*

---

[46]*The story of the development of this system of naming is comprehensively told by Michael Gordin in his 2015 book, Scientific Babel: How Science Was Done Before and After Global English.*

[47]*This is not to include the overlapping domains of the fragrance industry for which chemical knowledge is a primary source of business. IUPAC was created, in fact, to serve this very purpose of cross-disciplinary communication.*

Much of the standard chemical nomenclature reveals vestiges of its original folk taxonomies. At a time when the practices of chemistry were becoming more widespread, yet before the implementation of a universal nomenclature when behavioral characteristics grouped chemicals together nominally, the most representative chemical – either because of its particular behavior within the category, or because of less obvious things like its availability for laboratory use, or its smell – held siege over all of the other substance names, impressing yet more polysemous correlations on the map.

This prime representative of the folk taxonomy of chemical nomenclature is camphor. As a word, camphor follows a similar path to benzoin. It hails from Sanskrit, and it made its way into chemists' mouths because it was in high demand, not only for its delicious, purifying smell, but also for its more practical medicinal effects. Camphor occurs across many different species; even rosemary has a high concentration of the compound. It can also be artificially synthesized from alpha-pinene, a kind of turpentine (although this synthetic version is slightly different than the naturally occurring *D-camphor*). The natural camphor is most abundantly found under the genus *Cinnamomum* (in the species *camphora*). It is geographically limited to one small region, making it short in supply and thus a prime candidate to be synthesized. Thus, camphor has been sublimated, if you will, from its historical point of origin, to be deposited somewhere in the middle of Europe.

It was through experiments on camphor that chemists observed that some chemicals crystallize at room temperature after having been distilled. At this point the taxonomical position of camphor moved from subordinate to superordinate. Thus, mint, orris, and thyme are all referred to as camphor, because it is the most "common example" of this type of crystallized oil they create. The name-use of camphor is thus a common example of the tendencies of folk taxonomy to persist far enough into the era of scientific standardization to remain, to this day, as a source of potential misnomia.

Language of the Industry: Giving Form to the Ethereal

The language of chemistry serves the fragrance industry, of which there are others besides perfumery. These include the masking of bad odors, the scenting of artificially derived things that have no scent of their own, and the olfactive armament of cleaning products that relinquish the modern household from its daily pathogenic onslaught. Agronomists, botanists, natural products chemists, neurologists, quality-control analysts, and even lethal war gas designers all take part in the creation and use of odors for consumer use. These odor-related occupations not only are in charge of exploring, creating, and distributing smells, but they take part in the naming of those smells as well. Panels of "linguistic negotiations" are initiated for industry research (Holley, 2002: 20).The commodity of this industry, it seems, comes with no name, thus requiring the extra step before making contact with the public. The greater industry of smell is vast and timeless (note, the spices that set sail to Queen Isabella's ships), and perfumery is the dominant player. Being an artisanal craft and not so much a science, perfume is next in line for linguistic inspection.

The language of perfumery, through the lens of Modern English, began at a time when language, as well as everyday life, was less precise (Morris, 1984: 139). It was a time when intercontinental maritime trade routes performed the act of refining and reworking the names of the various substances traversing its network. The scent resides in the source, and, until very recently, it had been affixed to its biological source of origin, and, by extension, to its particular ecological niche, which may be, and frequently is, a unique place in the world. All of these places have their distinctive languages and their own names for the substances from which the fragrances are derived. These names are then exchanged amongst the traders, until the current English fragrance lexicon is formed.

This story is the same as the last, a semantic free-for-all in many circumstances. But perfumery is different from chemistry in one regard – perfumists need not adhere to a language intent on truth, or one that facilitates universal communication and considers potential permutations. For the Perfumist, the commerce of perfumery, and all it entails, is the driver of word creation and word use.[48] Many of the essential components, the basic palette of fragrance, were already established prior to or in simultaneity with the advent of chemistry. For this reason they retain their folkish flavor of non-scientific identification.

Perfumers, though distinctively different in their practical and linguistic use of the elements of smell, practice a method of identification and synthesis that is just as rigorous as that of the chemist. They may perform their penetrating analyses as a craft and not a science, but their acts are nonetheless as focused and as fecund as the other.[49]

On one end, perfumery is less concerned with the individual molecules than it is with their general category. These categories, called "accords," are similar to the way artists frequently refer to warm or cool colors. Accords are a form of soft identification: "floral" instead of "rose with a

---

[48]*Luca Turin presents this in his use of molecular vibration theory in fragrance chemistry, despite its controversial status amongst chemists. Scientific consensus or not, it works, from an industry standpoint, and so the problem is moot.*

[49]*This is in no way to discount the chemical prowess with which the perfumist must approach the craft. The point cannot be pressed enough in the procession of this text that chemistry and perfumery are less of a discriminable Venn diagram, and more of a circle, both overlapping.*

small bit of jasmine." They are intertwined in their intricacies, a movement of subdivision and superordination. Medicinal accords might refer to any number of incense, turpentine derivatives, etc. The Oriental blend permits enough room for interpretation, while providing at least some idea of the osphresiological whereabouts. Having considered already in some detail the opportunities for mistranslation, we can easily imagine situations in which it is more practical to speak in such broad strokes.

Calkin and Jellinek, in their crash-course text on perfumery, suggest a systematic process based on an understanding of how the human memory works, which they explain as using a network of autobiographical associations to conjure an artificial odor-object stimulus. They recount: "One team of perfumers were able to communicate among themselves the idea of a particular type of green note by the phrase 'smelling like elephant's feet'" (1994: 27-28). None of the crafters knows what elephant's feet smell like, or maybe they do. It is not about *veritas* – what the words really mean, to the "left brain," for example. Rather, it is what they evoke. This is a "right-brain" function that is at the core of the attempt to decrypt natural language in regards to smells.

Accords are thus the starting point for the perfumer's multipenetrating analysis of a complex fragrance. From the rough outline to the individual pixels, they move further down the resolution scale upon subsequent sniffs, revealing the constituent chemical signature, the Latin nomenclature. In the process of creation, they begin with skeletons of these accords, applying and removing sinew and muscle until the full body is formed (see Figure 7). There is a threshold to their dissection, however, which is based on the communicability of the consumer, the natural language of the general population.

Figure 7: The Structure of a Perfume. *Image found in Calkin & Jellinek (1994: 92), reproduced by Joe Scordo.*

The People's Language: Synesthesia and the Struggle for Verbalization

It is beyond mere trivia to note that regarding the consumer, the advertising of fragrance is an entirely synaesthetic affair. Upon reviewing the dynamics of olfaction as a neurological phenomenon, it becomes evident that the language of smell in the general population is like a Salvatore Dali painting – suspiciously vacant.

As they relate to the useful, universal scientific nomenclature, “Odors do not belong to classes determined by their chemical or physical attributes, [but] are coded in terms of sensations coinciding and interacting with them at higher centers in the nervous system” (Engen, 1982: 117). Those higher centers are the memory complex explored previously in this text. Because the identifiable semantic components of those memories will tend to be either auditory, visual, or tactile via the olfactory gatekeeper known as the limbic system, their reiterative existence as verbal exchange will tend towards sensory lexicons other than that of the nose.

For smell, people do not say “isovaleric acid with something pungent and bordering on putrid yet appetizing.” Rather, they say “smells like Parmesan.” The reason is because at some point in their life they had been in a similar situation, perhaps a fancy Italian restaurant or a *Nonna's* kitchen, and they are now recalling that episode. The *verification* of "Parmesan" does not come from the smell itself, which is the actual, physical substance evaporated into the air, and as a gas molecule nestling into the mucus surrounding your osmic brain-fingers. Rather, it arises from an associated episode where the same stimulus *was* verbally verified, in conjunction with other senses and a corroborating context.

Smell is contextual, and the habituating, categorical nature of the mind strings together a vast web of meanings and associations. “Whether or not a person can remember an odor's name (candle wax) or even an idiosyncratic verbal response to the odor (Sunday school class), does not capture the complete spectrum of semantic odor identification… (thoughts of certain people, visual images of certain places etc.) may not lend themselves to overt verbalization.” (Schab, 1995: 26) “The salient aspect of the sense of smell is the persistence of memories of episodes associated with odor…” and not memories of associated names of odors or other verbal response. Odor perception “is not organized lexically by sounds but around the similarity of objects causing odors and especially the contexts in which odors usually occur. A corollary is that they tend to be described in terms of personal preferences rather than by more general names like those proposed by traditional classification systems." The main function of smell is "not to recall odors for cognitive reasons but to respond to odors actually encountered" (Engen, 1987: 502-503).

The communication of olfactive sensory phenomenon, then, treats olfaction not as a perception *per se* but as a perception by proxy. In its place is the more abstractable, expressively manipulable sensory lexicon of sight, of sound, and especially of somatic perception. "No perfumer would describe a fragrance as 'methylphenylacetate-y' and certainly no writer of perfume ad copy..." (Morris, 1984: 228). Smells are warm, bright, or quiet, because the consumer is at a loss for making purely olfactory associations.

In synaesthetic delivery, Luca Turin comparts: coumarin – hay-like, nut-like, tobacco-like, "We are so used to associating this smell with buff-colored things that it is impossible not to think of the colour beige when smelling coumarin" (2006: 23). And again, on the animal component of para-cresol, castoreum, the mother of all beastly smells, "The very embodiment of the color brown in the realm of smell" (49). It is well-documented that the color of the substance emitting the odor will influence its description. As stated previously, white wines will taste red if they are colored red (Lawless, 1984).

As rational humans are wont to do with groupings that are too confusing to tabulate, the vast regions of the fragrance landscape are commonly arranged in a circle, or even a linear pattern for the purpose of layering the chromatic map atop.[50] "Florals" tend to fit within the reddish phase; "Animal" nests in the beige spot indicated by Turin; "Citrus" – could it *not* be yellow? – and "Green," well... .[51]

Overlapping as often as it does with the food industry, fragrance is often interpreted to the mind as taste. "Sweet" is frequently used as an olfactory descriptor. However, calling the caramel-smell sweet is not a chemical consequence of the molecule, but an effect of memory. If sucrose is paired with any odor that is smelled while at the same time tasting something sweet, then the odor will eventually "smell sweet," meaning it will be perceived as sweet, although nothing about the odor itself has changed. The potential effect on odor perception that is cued by such multisensory information is thought to be the minimization of "perceptual dissonance between the dominant sense (vision) and the minor sense" (Wilson & Stevenson, 2006: 209).

---

[50] *See: Aftelier's Natural Fragrance Chart.*

[51] *"Green" is a fragrance category represented by "freshly mown grass," as well as by the molecule cis-3-hexenal (leaf aldehyde).*

Smells are not just sweet; bacon is umami, and body odor, sour. Nevertheless, the word, the taste, and the smell of "sweet" are fundamentally linked. This link can be evidenced in the fact that the word "sweet" is the most commonly used subcategory heading in the Aldrich Catalog (2011), which is a chemical clearinghouse for flavor and fragrance scientists. The biases and idiosyncrasies inherent in the Aldrich Catalog as a universal *aroma* reference are immediate and comprehensive. The catalog is, first of all, a flavor tool. As a commercially driven pursuit, flavor reigns supreme over fragrance alone. The fact that "Sweet" as a subcategory has the most centrality – it falls under Balsamic, Woody, Fruity, and Minty – could relate to a confluence of the pervasive availability of sweet aromas or to the sweet tooth of a certain society. Alternatively, there really could be something sensually similar about the relative phenomena. Unresolved in that matter, we can scrape together at least that the description of a smell will tend to be communicated via other, more intentionally simulable channels of association – anything but the olfactory-specific semantic network of experience. The reason, of course, is that such a thing is so halting and reluctant in its cooperation, if at all.

In parallel to this gustatory congruency of sweetness, it is a curious thing to consider: Are florals such an extensive part of the perfumer's repertoire because of their associated *visual* aesthetic beauty?[52,53] David Howes (2002) responds by recounting his work with the Kwoma people of Papua New Guinea. Howes presents plastic cards impregnated with scent and painted with a corresponding color (cinnamon is brown, coconut is white, etc.). When surveyed on the most-liked odor, the vast majority of respondents voted for the rose sample. When Howes put the red-colored rose scent in an envelope hidden from view, however, the overwhelming preference disappeared (along with Howes's search for a universal aroma preference, à la Berlin and Kay, 1969).

Building the Phantom Lexicon

In the spirit of the renowned physicist and thinker Richard Feynman, let us recall "What I cannot build, I cannot understand." A thorough understanding of our sense of smell, or of anything for that matter, requires

---

[52]*According to the Aftelier chart for natural fragrances (2013), Floral has double the subdescriptors as compared to all others.*

[53]*Aside; Floral is the only main category in the Aldrich catalog that does not connect via its subcategories to any other. The subcategory Sweet shows up under Balsamic, Fruity, Minty, and Woody. The subcategories of Floral, such as Geranium, Rose, Jasmine, etc., are not duplicated under any other main category. Floral is the only one to do that. Sweetness (most central) and Floral (most isolated), therefore, exhibit a curious kind of polarization of the aromasphere; at least as far as the Flavor/Fragrance domain of the specified catalog is concerned.*

a set of defined terms. Precision of language is the basis upon which knowledge is availed. With no precise words to fill our mouths, our thoughts are thwarted at the outset.

In having sketched the structure of smell's lexicon, it has not been my intent to prove that such a lexicon cannot be built. Rather, my contention is that that when the lexicon is constructed, it must be done very carefully, and with an expectation not of the absolute, but of the approximate. This kind of information environment poses a unique set of problems. Each variable, or word of the lexicon, is a range of probability and of stipulated disambiguation. What results is not a clear picture, but a map nonetheless. To the main features of this landscape the text now turns.

6. The Classification of Smell

Much has already been said or implied about the organization of this "language" of smell. Both the meanderings of history and the first attempts at clarity by way of chemistry have donated to its chimerical formation. One point carries over from the previous chapter: If smell is so resistant to naming, then it is certainly resistant to classification. The basis of such a taxonomy is a prior semantic web. However, semanticity, the naming of things, is by its nature a cognitive act. Olfaction is *not* a cognitive act, and therefore it is not given much to the practice of semanticization. This reality creates less of a taxonomy and more of a list; in fact, an infinity of lists, or a Celestial Emporium, at that.[54] For the purpose of imposing a basic outline, these lists will limit the kinds of classifications to the odor as a feeling, then as a context, and finally as a source.

As a Feeling

Smell is a verb, implying a subjective actor. The act of olfaction, although it functions unconsciously most of the time, is a rich experience, a multifaceted phenomenon. At the center of that phenomenon is the subject, and the preconscious, protosensory interface, that of emotion, or feeling (technically, emotions are more like feelings *of feelings*; see Damasio, 2010). Emotions occur on a continuum of pleasure-pain, a spectrum previously mentioned as the hedonic scale.

The most simple, straightforward, and commonsense approach to the organization of smell should be to assign it a hedonic value: It is Good or Bad. After all, olfactory stimuli either repel or attract the organism; the *limbic* system controls the *limbs*, and thus the means of mobility. Olfaction is thus a primitive motility actuator, an incipient cognition instigating the mobile organism. As reviewed previously, the majority of the neural activity involved in olfaction is dedicated to managing the emotional reaction to an odor (and its subsequent memory correlates). In turn, this serves as a means to direct the course of action for the organism. This reaction is expressed in the hedonic dyad, the one used by Plato in his *Timaeus*, where he refuses to confine smells to different kinds. They have

---

[54]*Reference to writer and polymath Umberto Eco's Infinity of Lists, 2009, which is based on the imagery in Achilles' shield as described in the Iliad, a finite object representing the infinite; and Jorge Luis Borges' 1942 description of a Celestial Emporium meant to express the arbitrary and absurd attempt to create a systematic classification of everything.*

no names, he contends, and can be distinguished only as painful and pleasant (1961: 1190).

Identifying hedonic value would insinuate the naming of the corresponding source of the stimulus. The two are, on the contrary, independent of each other, as it is generally understood that there is no universality in the hedonic value of a particular smell. Human olfaction is not hardwired. Just the opposite: It is learned, and different people, for various reasons, learn about smells differently. Imagine, for a moment, the Indus Valley civilization. A large portion of it lies below the water table. Consequently, bodies, *the dead*, are not buried underground; instead, they are burned above it. To the inhabitants the cremation might smell like incense, or a grandmother's perfume – like nostalgia. The smell of burning flesh means something very different to these people than it does to people who bury their dead underground (to burn much more slowly in the fires of bacterial decomposition).

These examples of cultural differences are abounding in the olfactory research (though it is not always made explicit). Two studies emphasize this observation. In the first study, wintergreen (methyl salicylate) was reported as pleasant by Americans, who associated it with candy, but unpleasant for British subjects, for whom it was a reminder of the customary pain reliever of their time (Moncrieff, 1966; Cain & Johnson, 1978). In the second study, a demographic of Germans (unknowingly) reported that ketchup flavored with vanilla was more pleasant than "regular" ketchup. The author of the study attributed this strange preference to the participants' exposure to vanilla-flavored breastmilk, which was popular in Germany at the time (Haller et al., 1999). Such factors of cultural contingency have made the mapping of hedonic value-to-source a fruitless effort thus far.

There is one (and only one) exception to this status of the non-universality of hedonic value in odor substances; it is eerily immortalized in the opening passages of Patrick Süskind's novel *Perfume* (1986). The main character is yet a baby, an orphan, rejected by his wet-nurse due to his freakish absence of bodily odors, especially from the top of his head, "where they smell best of all." In the mutually reinforcing *yin-yang* of post-pheromonic olfactive interchange, the mother's nipple is to the baby's head, *ad infinitum*. It is thus stated that there is only one smell that is universally recognized as pleasant, regardless of culture, and it is the head of a baby, which smells like a kind of sweet buttermilk perhaps, and as an extension of the pan-human baby diet of breastmilk.

The hedonic dimension is asymmetrical, weighing much more heavily on the negative allocation. Following a series of experimental endeavors, a clear picture of the role of the negative or unpleasant side of smell emerged. It appears that olfaction – as could be posited for all senses –

gives priority to stimuli that may be harmful (Mouélé et al., 1997). Negative olfactory hedonics is the most physiologically "visible," showing increased heart rate and activation of the amygdala in response to unpleasant odors. In contrast, neutral and pleasant odors exhibit dormancy (Alaoui-Ismaïli et al., 1997; Zald & Pardo, 1997, resp.). It is also the most nominally visible – bad smells occupy more of the odor namespace than any other distinction, making the hedonic taxonomy quite phobiaphilic (Boisson, 1997). Finally, some odor specialists postulate that such negatively valenced stimuli are processed in two neurologically distinct ways: a "quick and dirty" route for potentially harmful substances and a more "cognitively complex" route for pleasant/neutral odors (Rouby & Bensafi, 2002: 154-55).

There is one study, however, that seems to contradict this apparent trend in behavior. In fact, upon closer examination, it reveals the more nuanced dynamics of the olfactory system. In an attempt to map the semantic web of *subjective* responses to smells, Chrea et al. (2009) reported that the vocabulary for pleasant odors outweighs that for unpleasant odors. Note that this experiment was designed to maximize emotional smells, and thus more *subjective* responses. To ensure this, the scientists chose smells that would be familiar to the subject, thus increasing the likelihood of an autobiographical experience. Herein lies the problem: Odors are limbic, there is no cognition available. The only language of the limbic system is fear.[55] "Bad" smells and unfamiliar odors have less of a vocabulary to name them because they are not handled by the linguistically rhythmed mind. Instead, they function deep within the nervous system and do not reach into the naming brain. There is no time for contemplation, only reaction. Only after subsequent interactions of positively valenced familiarization can a smell develop a name, or even an adjectival description.

Research has also revealed that there is more agreement across cultures as to what "bad smells" are and, conversely, more variability as to good or neutral odors (Schaal et al., 1998). In a way this observation is in line with the Rouby and Bensafi hypothesis cited above. Smell initially shows up on the linguistic radar only when it is bad. Thereafter it appears via the complex mess of cultural mediation. As more odors are named, the new ones can move across the spectrum.

The culturally specific subject has entered as a most important factor in determining hedonic value. This dyad, with its two values, can be interpreted as a continuum in the way color names are separated. There are varying gradations of Unpleasant, from annoying to suffocating. This is the

---

[55] *In the case of olfaction, this fear comes in the form of disgust (except in the case of environmental hazards where it is simply "fear"). In this very study, it was found that the negatively valenced terms revolved around one emotion – disgust. Note here the link between olfaction and taste (gustation).*

general idea of using a hedonic value to measure and categorize smells. When we consider the shifting state-matrices of such taxonomy as a function of its source culture, it functions more as an indicator of cultural characteristics than as a means for an articulated body of odors. Hedonics is now fully decoupled from its odorant sources by way of this contextual intermediary. It is towards this greater context that the classificatory attempt now wanders.

As Contextually Embedded

A brief review of some of the most salient features of the phenomenon of smell is appropriate. Smelling is a predominantly unconscious activity, running below the level of cognitive oversight but within the body-minded limbic system. The interoceptive limbic states elicited by olfactory stimuli inform complex arrangements of episodic memory. This process has been agreed upon by scientists and scholars for some time – we do not smell individual components, but the whole.[56] Further, it is not only the individual chemical components that are lumped together. Instead, it is this complete odor *as well as* the context in which it is embedded. The episode *is* the identity, and it is inextricable from the odor (lest it lose all of its meaning!).

Olfactive perception is a galvanizing experience, and once this embedding has been initiated, there is no disintegration. It is because of its "intimate integration," Engen contends, that odor classification has contributed nothing to the sense of smell (1991: xi). Insomuch as a general contextualization yields semantic identities such as "rain on warm asphalt," the autobiographical nature of smell grows names like "Christmas at Uncle Clark's." This leads the way to the truly Borgesian taxonomy. Places; people; things on fire; things warm; things soaking wet; times of year; stages of fermentation; stages of decay; all diseases that have an identifiable corresponding odor; new cotton shirts and cotton shirts worn by either men, women, children, or elderly people; smells as they occur during menstruation; chicken soup when healthy, chicken soup when sick (just kidding – there is no smelling with a head cold); a cigarette after one beer, and a cigarette after 12. Space, time, phylogeny and ontogeny, and thus even the upper taxons have no reasonable "end" in such a high-density associative network. Upon smelling a coconut-scented plastic card, the Kwoma people of Papua New Guinea (mentioned previously, in Chapter 5) brought back coconut-flavored cookies from the trading post rather than

[56]*See one of the most important works on olfaction in recent history (for which a Nobel Prize was awarded) on "White Smell" (Weiss et al., 2012), but more important the information-processing techniques referenced therein, which all make use of the idea of odor-identification as a gestalt process, not an analytical one.*

the actual coconuts that grew plentifully in the area (Howes, 2002: 76). Veridicality, or semantic truth, is fleeting in such subjects.

Even the sequential context of odors in a seemingly "clean" test experiment can inadvertently bias identification by context-dependent associations. For example, it has been observed that learning an odor is enhanced if a color is presented before the odor, but not after. This finding suggests that cue order has a major impact due to the ecology of odor perception. Because odorants are normally associated with the source from which they emanate (cue, then odor), we identify the source first and *then* associate it with the odor. But, "When an odorant comes first, resources are devoted to generating its name *at the cost of paying attention* to the stimulus that follows" (Wilson & Stevenson, 2006: 201, italics added).

How can we reconcile this opposition between smell as contextual and classification as semantically driven? Olfactory scientists refer to "odorants" as the chemicals themselves, and "odors" as the resulting perception. Much semantic leeway has been afforded by the classifying structures of Danièle Dubois and her colleagues. Dubois specializes in cognitive psychology and language, and she provides an alternative for the rigid semantic nets used to conduct scientific research on smell (2007). Dubois suggests that olfactory research hesitates between "investigation about *responses to odorants* (physical substances) and *representation of odors* (psychological concept)..." (p177).[57] She then goes on to establish an opposition of lists, of physical effects versus semantic labels – of *answers* versus *names*.[58]

Instead of following the traditional protocol of establishing a single name as an answer (Apple), Dubois uses an answer-group. This group includes specifying adjectives (green apple), source-relative adjectives (apple-like), and source-related artifacts (apple shampoo). The potential answer then branches out to the more generic "answers" that relate the subject's *experience* with the odor (fruit, food, candy, cleaning spray, etc.). Next in the expanding sphere of answer groups are the co-homonyms, (therein, apple is named by participants as strawberry, raspberry, and lime).[59] Orange and Lemon are co-homonyms, and one is often mistaken for the other. Co-homonyms can form as a result of a shared superordinate such as Fruit. Ultimately, however, they are determined by the particular corpus of reported answers – for whatever reason, Lime is a co-homonym of Apple in the study.[60] Next, a second-level co-homonym mistakes the superordinate category, as in calling Apple "Flower" instead of "Fruity."

---

[57]*Dubois herself is citing Hudson and Distell, 2002.*
[58]*Dubois references a more formal study; see Dubois & Rouby, 2002.*
[59]*These were reported from the Dubois & Rouby 2002 study and are not necessarily representative of general relationships between odors and subjects.*
[60]*Although a second thought should prompt one to consider the aforementioned effect of the sequence of odors presented. If Apple were to come after Lemon – Lemon being a typically*

From here, the list goes on to contain such things as the deverbal, adjoining hedonic values introduced earlier (nauseating, agreeable) and Engen's "tip of the nose" answers. The latter are akin to giving no *nominal* answer at all: "I've smelled this before, but...I can't remember...." An interesting distinction occurs here, which is that adding *–able* to one of these words (pleasur-able, disagree-able) indicates a possibility, whereas adding *–ant* (pleas-ant, repugn-ant) constructs a past tense. This phenomenon suggests an actual effect, and hence the subjective nature of the smell-language.[61]

A final articulation, which expands this potential answer group to the greater psychological and social considerations: Prospective answers can use "personal marks" – words such as *my* and *us* – to distinguish between objective utterances and subjective ones, as well as between collective references and individual ones.[62] Definitions in general are more frequently given as collective, because mutual agreement or consensus is inherent in the idea of a definition. Thus, Dubois asserts that what is *reported* (in light of these personal marks) about the odor is not the subjective experience *as it occurs to the subject.* Rather, it is the experience as it is *collectively known to be* on a personal level. This is an important distinction because smell is such a subjectively informed phenomenon. This intersubjectivity makes smell naming void of the social trust enabled by the equalizing medium of the distal senses (Köster, 2002: 29). Grossen labels this process "an illusion grounded in consensus" (1989: 52).

Before the recursive loop gains strength, the discussion must conclude and change course. Using a more robust structure of smell identification such as Dubois and colleagues' "answer groups," a contextual taxonomy could then be more easily mapped with the hedonic values of specific demographics just discussed. Yet still, a practical, functioning taxonomy of smell has been avoided. To the last and final attempt the discussion proceeds.

"easy" target, and Apple not – then Lime might be a good "guess." Kurtz et al. (2000) found that similarity ratings changed with substitution of one odor in the sequence.

[61]*Note, this study was conducted entirely in French.*

[62]*This in particular is referring to Sophie David, 1997.*

As a Source

"The only way to classify smells is by their sources."
-Dan Sperber, Cognitive Social Scientist, 1975

"Smells have no autonomy vis-à-vis their sources."
-Sophie David, Linguist, 2002

Darwin, in designing his tree of evolution, used as a template the existing maps of languages as they were arranged across time and throughout history. The Linnaean biological taxonomy existed at the time, but its organization was not conducive to expressing the idea of evolution. So, Darwin used the genealogy of language instead (1859). Here the discussion picks up, knowing full well that the relationships amongst the organic sources of smells are a poor template for organizing a meaningful taxonomy for the aromas they afford, because the language of smell evolves both throughout history and within the individual. Moving from the subjective spectrum of hedonic valence, past the omnicategorical net of contextual identity, the path of discourse now takes odor to its objective molecular limit. To do this we must address its source.

Evaporating Organic Matter

Because negative odors are more prevalent in our vocabulary, the list of dead, decaying, rotten, sour, moldy odors in the everyday lives of all people must be outlined forthright. These are smells that have been managed in different ways by different groups of people but that have yet to be rid of completely. Organisms die and decompose, and in the process – a heated process aided by bacterial breakdown – they vaporize aromatic warnings in communication to the living. Their molecular interosphere comprises a laboratory of chemical reactions; they exhale noxious fumes with names such as cadaverine and putrescine. Other analogs to such heated decomposition are the more extreme version of burnt things and the more placid version of mold growth, both of which, like the smell of rotten flesh, are important signals of their associated negative *causes*. Even miasmic exhalations of the earth itself in the form of methane belches, for example, are warnings of impending danger.[63]

Human body odors make up the next major category of (potentially) negative smells.[64] What would typically be called "body odor" is not so

---

[63]*Sometimes, however, such vapors from the bowels of the earth are odorless, and they have been responsible for many incidents of instantaneous mass deaths. In fact, many gases toxic to humans are odorless "silent killers."*

much a human-made odor as a bacterial one. As sweat evaporates from particular regions of the body (armpits, genitals), it leaves behind a mixture of substances on the surface that are then eaten by bacteria. The resulting bacterial metabolism generates most of what is considered body odor. Because young people have not yet had their bodies colonized by bacteria, their odor is less intense. By contrast, elderly people have a distinct scent, which is commonly called "old people smell," reminiscent of "old book smell," and nominally noted as trans-2-nonenal (a major constituent of the odor). Finally, regarding human body odors, there is the smell of disease and the dying. Anecdotally, it is said that upon having to fastidiously relocate an entire early 20th century cemetery, workers could tell the differences among influenza, diphtheria, tuberculosis, and typhoid *in the soil itself* after 100 years of having bodies buried in it (Protectedstatic, 2012). Non-anecdotally, many diseases have distinctive corresponding smells, which are still used today to aid medical practitioners. Even mental illness in the form of schizophrenia has its own scent.[65]

There is an overlap here, between this initial category and what seems to be one of the only usable examples of a common lexical organization for odors. In an attempt at olfactory ethnography, cognitive neuroscientists performed an inquiry in 60 languages across nine language families. The data reads as follows: Of all the languages, 35 have specific names for sweat/body odor, 34 for strong animal odors, etc. Overall there were four groups of odor names – body odors, decaying organic substrates, animal foods, and live animals, with a heavy reliance on the "burnt, rotten, rancid, moldy" vapors that populate this negatively valenced field (Boisson, 1997; revisited by Rouby & Bensafi, 2002: 149 Table 9.2). (Note: Names for "feces" show up in only four languages, prompting Boisson to question whether this is a taboo that is too strong to mention.)

By way of the aforementioned groupings, it is abundantly clear that the anthropogenic smellscape is concerned not so much with the source of the smell *per se* but, rather, with a process, a hidden metabolism at work that makes itself visible through such exhalations. Plato describes odor as emanating from a body undergoing change, eliciting an intermediate state of instability (1961: 1190). This is also a reminder that the negative pole of olfaction is the one that is "linguistically marked" (Mouélé et al., 1997). As for the rest of the hedonic spectrum – those odors more commonly

---

[64]*It should be noted that these human-generated smells are not necessarily negative. The fuming body odor of a loved one is not the same as that of a stranger. (This articulation can be extrapolated to larger social behaviors as well, evidenced in the history of olfactory-based prejudices against "the other.")*

[65]*This point is fantastically queried by the odor author Annick Le Guérer (1992), where she plays with the possibility of the "odor of sanctum" reported to emanate from certain saintly corpses as a result of extensive abnormal mental states which lower, or encumber the metabolic rate, leading to incomplete combustion of aromatic materials in the body. She reciprocates by suggesting such lower metabolism as a result of sustained meditation. Regardless, it is a general understanding that psychosis brings with it an identifiable smell.*

recognized as “fragrant” – they tend to come from plants, not from people or detritivores.

Botanical Organisms

Aside from botanical sources, there are animal sources for the odors of civet (cat), ambergris (whale), musk (deer), and castoreum (beaver). In addition, there are the synthetic sources that are produced in a laboratory even from nonorganic chemicals. Significantly, some of these chemicals identically represent their naturally occurring odor twins, whereas others smell like nothing of known natural origin. These animal sources are rare, and, due to the discovery of synthetic versions, in combination with ethical issues, their use has become less common (even if the desire for their intoxicating effects has not).

Not only are many aromatic botanical sources re-created, or synthesized, out of chemicals that are sometimes unrelated, but this process often occurs in reverse: Individual monomolecular odors are separated from the overall grouping and treated as representative of the “natural” odor. Still, the botanical instantiation of the odor is a natural locus, especially because it carries with it such multimodal, contextual information. This information is both episodic and semantic in its associations; both types are necessary for the cognitive endeavor of organization.

All plants have a binary Latin name that serves to differentiate between *Cinnamomum cassia* and *Cinnamomum verum*, for example, which otherwise would both be called “cinnamon.” Although cinnamon is basically limited to modern-day Sri Lanka, other plants must have their origin specified; for example, the lemon oils of California and of Argentina are distinguished from each other (Aldrich, 2011). Each plant can yield different fragrances depending on which part of the plant is used. Barks, flowers, fruits, leaves, resins, roots, seeds, and woods of the plant can all produce aromatic compounds, and two different parts of the same plant can have two different smells. The coriander seed and cilantro leaves, both of the *Coriandrum sativum*, are so different in their aroma that they have taken on separate names. Usually there is no distinction as to the part of the plant. Moreover, at times a given name can actually be *misinformative*. Crossing the chemical threshold for a moment, flowers of benzoin are a beautiful example of such misnomia because they are actually crystals, not flowers.

In that segue, most botanical sources can be concentrated through laboratory mechanisms into essential oils or the like, through processes of distilling, dry-heating, pressing, and extraction through the use of solvents. And, of course, different extraction methods create different smells, so much so that orange blossoms obtained through solvent extraction are called "orange blossom absolute," whereas the same ingredient obtained through steam distillation is called "neroli oil."At this point there are a handful of subcategories to be addressed in properly identifying the botanical source: the specific plant anatomy being used, the place of origin, and the process of extraction.

Chemical Products

Chemicals themselves are grouped into families based on their structure or the way they react with other chemicals. Esters, for example, result from a reaction between an alcohol and a carboxylic acid. Aldehydes have a double bond between oxygen and hydrogen atoms. Note, however, that besides being loosely organized on the basis of the volatility of organic chemistry, these categories are distorted upon transitioning into other fields. Perfumists may refer to a vanilla scent as aldehydic, though vanillin is not an aldehyde.

As the pattern requires, the classification of organic chemistry follows many rules simultaneously. Prior to the full implementation of organic chemistry, smells were referred to by their general sources, with the aforementioned subcategories being amended as needed. In some cases, such as benzoin and benzoic acid, the specific representative monomolecule is named after its aroma-associated source. As for isovaleric acid and trimethylamine, their names are almost entirely chemistry-generated.[66] This bifurcation opens the next system of odor organization, one that uses laboratory methods of induced chemical reactions, as well as molecular structure and composition, to generate the subclassifications of odors *as mediated by* the science of (organic) chemistry.[67]

"Why should smoke possess only the name 'smoke,' when from minute to minute, second to second, the amalgam of hundreds of odors mixed iridescently into ever new and changing unities," Grenouille asks in Süskind's *Perfume* (1986: 25). Granted, there is a supernaturally hyperosmic lunatic speaking here, but he has a point. Smells, as they occur naturally, are rarely a single odor molecule. Instead they are an odor profile – a distribution of the amounts of different chemicals.[68] Were they to be artificially isolated, or even left on their own, their state might be subject to change depending on the conditions. Normal amounts of oxygen in the air

---

[66] *As with many common odor chemicals, isovaleric acid is the common name, 3-Methylbutanoic acid the IUPAC name. It is the smell of sweaty feet, vomit, horribly rancid milk, and Parmesan cheese. Trimethylamine, or N,N-dimethylmethanamin, is the smell of rotting fish.*

[67] *While the general population seems to arrange their lexicon according to the negative smells, and fragrance-smiths have developed the vast measure of the olfactory map according to the pleasant ones, the pursuit of chemical knowledge has been, for the most part, impartial to the hedonic valence of the smells that emanate from its laboratories. This being said, it is interesting, or obvious, that because of the timelessness of perfumery, the names of positively valenced smells are of "Perfume"-origin, leaving many of the bad smells to be only later named by "Chemistry," thus delivering a lexically interchangeable taxonomy.*

[68] *A primary constituent of wood smoke is called guaiacol.*

are enough to change a sensitive odor molecule into something very different; such reactions are a major consideration in food design and artificial flavoring. Terpenes, which comprise an important element of citrus-smell, oxidize over time, changing the relative proportion of the overall mixture, and moving from "citrus" to "turpentine." Moving in the opposite direction, butyric acid, which smells like the rancid butter from which its name is rendered, can react with ethanol to become ethyl butyrate, which smells of pineapple. Ethyl butyrate is then used as a primary ingredient in artificial orange juice flavor. Only with careful consideration can an odor be named with any degree of accuracy. The name type is just as important: Is this a reference to a specific chemical within the orange profile, or to the entire smell-of-orange as it occurs in the pureness of its essential oil? In the case of the ethyl butyrate and orange juice connection, the former is synthetically derived and thus does not actually exist as a molecule in the orange (Sigma-Aldrich, 2013).

Herein, when referred to as chemical products, smell-names and their ensuing organization are comprised of monomolecules. It is via the mystical transmutation of chemical synthesis that these odor molecules carry with them confusing links to other, seemingly unrelated odor molecules. Some examples of this phenomenon have been visited previously in this text, such as the case of "benzoin" (as a greater taxon, or lexical unit). The chemical product benzoic acid can be acquired not only from the botanical specimen of the genus *Styrax* but from the bitter almond as well, and through a variety of processes.[69] Benzaldehyde smells of bitter almond extract – it is often referred to as the smell of cyanide, or poison – and it is incontestably different from the resinous, camphoraceous smell of the gum of benzoin. Yet from a synthetic odorcentric perspective, benzoin resin and bitter almonds are related. Any chance at a hierarchical organizing structure is totally encumbered by the resulting relationships between naturally occurring odors and their synthetic pseudo-twins.

In fact, due to the historical significance of benzene, or "aromatic benzoin," this omnipresent lexical entity not only refers to the nominally related chemicals listed above, but it is a taxon that encompasses the entirety of organic chemistry. In other words, to consider odor molecules as chemical products is to reveal multiplicative similarities that have little to do with odor characteristics and yet are required for clarifying ambiguity and misinformation. Take, for example, natural gas, which is odorless. In order to make it detectable, a unique skunk-like odor called ethyl mercaptan is added. In this case, the parts are affixed to each other by way of chemistry as well as by the greater modern enterprise of civil engineering.[70]

---

[69] *The bitter almond (Prunus dulcis var. amara), in opposition to the sweet almond (Prunus dulcis var. dulcis), yields a large amount of benzaldehyde, from which the chemical benzoic acid can be synthesized. Benzoic acid is the representative odor monomolecule of benzoin resin.*

On the topic of misinformation, and with such talk of monomolecules and representative constituents, the olfactory tendency to redintegration must be addressed. Part of a smell can carry with it the co-occurring odor molecules around them in the memory, and it will later be used to substitute for the whole. Strains of cannabis, aside from the strong skunk-like smell, can have significant amounts of limonene in them. Through redintegration, the potent smells of such cannabis become so tied together that upon smelling an orange (almost entirely limonene), a frequent user might hallucinate the other odors of cannabis along with the orange. This phenomenon represents an apparition superimposed in order to satisfy the nose-brain's insistence on predicting an odor based on limited or partial information – a behavior that is not limited to olfactory perception.

One of the few distinct reverse instances of the chemical naming of the scent over the botanical is that of the synthetically derived "Marine" class of odors, "oceanic," "watery," the scent of the seashore: It is a secondary metabolite of seaweed pheromones. The original synthetic version of this scent was discovered by a pharmaceutical company and subsequently refined by a fragrance company. Although this naturally derived form is present in brown algae, it is not used in perfumery, at least in the Western tradition. The surest evidence of this is the simple fact that this synthetic aroma created an entirely new branch of Modern perfumery in the way the "benzene ring" created organic chemistry.[71]

The advent of, and the interplay between commercial entities signals a new shift in the organization of chemistry-originating odor molecules. Enter the most triumphant of synthetic odors – musk – called white musk to differentiate from the natural version, which is extracted from a species of deer. Musk scent is desired the world over. However, it occurs naturally only as a bodily fluid of an animal, making it both prohibitively expensive and ethically dubious. Within this section it is mentioned on account of its (accidental) discovery and its subsequent incidental olfactory relations. Chemist Albert Baur was conducting experimental research on TNT, when he noticed that some of the nitrated derivatives of benzene smelled pleasant (Mookherjee & Wilson, 1982). This observation prompted the first synthetic musk family, called the nitro musks. Once again by way of chemical synthesis, two seemingly unrelated odors – synthetic musk and organic benzoin resin – are cryptically related, requiring an organization that is anything but clean cladistics.

---

[70] *The history of civil engineering, interlaced with urban planning, is parallel to that of olfactory awareness, at least through the lens of French society circa the 1700s. See: The Foul and the Fragrant, Corbin (1986).*

[71] *The trade name for this molecule is Calone, and the chemical name is methylbenzodioxepinone. The pharmaceutical company that discovered the original "marine" molecule was researching antidepressants, known as benzodiazepines. (Note the omnipresence of Aromatic Benzene.)*

The story of white musk continues. Being related to the dangerous substance TNT, such "nitro musks" were left behind in favor of a new synthetic process and its new imitative class called the nitro-free, polycyclic musks. This time, the fragrance industry created the imitation on purpose, instead of simply borrowing from the mistakes of science. In the manner of the commercial entity, it begot a tradename – Phantolide (a registered trademark). Significantly, the musk's initial semantic identity was not generated by chemists so much as by commerce. In fact, the chemical itself, acetyl indane, was unknown to chemistry proper until a few years after its discovery (Kraft, 2004: 152). The broken loop of proprietary information and public knowledge in the realm of odorcentric communication thus demands a new layer, another double-articulation to be addressed.

Exhausted and yet incomplete, this concludes the discourse over the ways in which smells are broken down from the aggregate and reassigned within a given paradigm of configuration. The taxonomies generated from these different modes of organization are to follow.

Taxons Proper

There are three fundamental versions of odor-object taxonomy. One is created strictly for organizational purposes; it falls within the domain of science. Another exists for its usefulness in identifying and creating artificially constructed fragrances; it is subject to the more nebulous realms of art.[72] The final taxonomy is one created, albeit semi-unconsciously, by and for the people.

The Science

The same scientist to whom botanical classification is attributed is also the one who recorded the first attempt at classifying odors (and not by coincidence, because odors are primarily a botanical phenomenon). Around 1750, Linnaeus presented roughly seven odors following from pleasant floral to neutral garlic/goat, to putrid rotting fish. Notice the limiting number here, as seven seems to be the natural denomination for sensory categorization (seven colors, seven tones, and George Miller's

---

[72] *This is to the benefit of Chemistry, and not olfaction necessarily. The naming and organizing of chemicals is done independently of their odors, and many of the odor characteristics of chemicals are discovered by accident. It is because of the disconnect between chemical phenomena and olfactory phenomena that odors are only used as descriptors of chemicals, but not as a means for organizing them. In the circumstance where odor does seem to play an associative-determinant, such as with Benzene, it is, or at least regarding the purpose of this text should be considered, incidental.*

chunking-based mnemonic theory, which explains why phone numbers are comprised of seven digits). Surely the primary categories listed by Linnaeus and subsequent scientists were not chosen at random. In fact, they follow the hedonic spectrum, in which each odor functions as the primary representative of a particular shade of affect.[73] Yet, upon further extrapolation, it becomes obvious that they do not give rise to the rest of the olfactory inventory. This is not due to sheer volume, of course. The *Systema Naturae* itself (Linnaeus' naturalistic tome) contains a mountain of entities. Rather, it reflects our inability to deconstruct our olfactory perceptions in a way that allows meaningful comparison.[74]

The history of classification follows thusly from Linnaeus in ever-expanding groupings and variations of such "primary" odors, in the way that all colors are derived from the three primaries: red, yellow, and blue. Next, psychologist Hans Henning added a three-dimensional model to the heretofore semantically driven operation (1916). Henning posited six odors – flowery, fruity, spicy, resinous, burned, and putrid – and placed them at the vertices of a triangular prism (see Figure 8). Any odor should then find a place *in-between* these vertices (on the surface of the prism, not within its space).

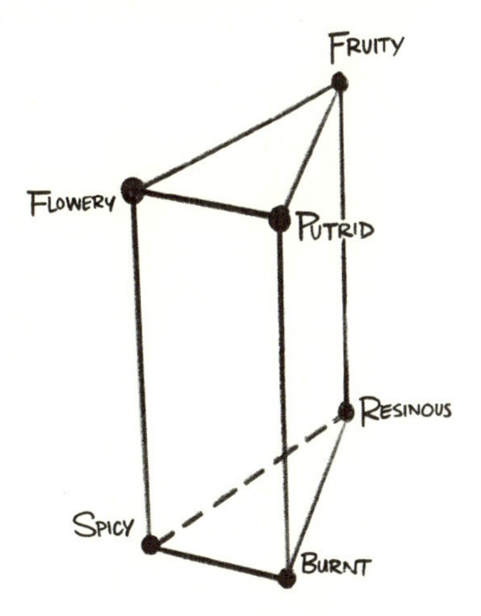

Figure 8: Henning's Smell Prism: The three-dimensional smell space of German psychologist Hans Henning, 1916. Imagine the smell of apple being somewhere between "flowery" and "fruity," until it begins to rot, at which point it moves towards "putrid." *Illustration by Joe Scordo.*

---

[73] *That one particular odor may be more of a representative over another could be either an indication of that odor's intensity, or a cultural ubiquity/familiarity.*

[74] *The fact that Linnaeus was concerned with odors as an indicator of a plant's medicinal use should be noted.*

Besides simply adding more primaries to the list and allowing overlaps or different interstitial-dimensional values between them, as most proceeding attempts continued to do, a novel approach came out of developments in genetics, neurology, and olfactive perception. Biology informs us that the nose contains roughly 400 receptor neurons that are used to encode the infinite odor catalogue. Further, each of these receptors is "controlled by" its own gene, making the olfactory apparatus the largest gene family in the human genome. Many of the known genes for human olfactory reception are actually inactive and thus are called nonfunctional pseudogenes.[75] (In dolphins, all of their olfactory genes are junk; dolphins can't smell.)[76] It is quite common for one of these receptors to be in poor working condition. This scenario causes specific anosmia, or the inability to smell a particular odor, such as asparagus-urine (sotolon), seminal fluid (putrescine), or flatulence (hydrogen sulfide). Biochemist John Amoore collected all 40 known anosmias, thereby creating the first receptor-centric list of anosmic primaries (1969). Again, however, when scientists attempt to elicit even a short description of each of these anosmic-primary odors, the overlaps are confusing, rendering the list useless for practical purposes. Unfortunately, scientists have been unable to make sense of the olfactory rainbow. The very concept of primary odors ended with Amoore's studies.

With the increased use of statistical probability and the rise of information science as a means of clarifying such noisy data as this olfactive endeavor, the classification of odors began to reveal its contextual sensitivity. Avoiding previously determined "correct answers," or taxons, the dataset itself can suggest the proper organization. Using such a statistical method, new odor investigators collect large sets. One prominent example is Arctander's 1969 *Perfume and Fragrance Chemicals*, which lists more than 1000 chemicals and corresponding descriptors. The descriptions, or "notes," are plotted against the chemical compounds, the unique odorific monomolecules. What is then measured is the *distance between notes* for each molecule. Each odor is then described as a position in an *n*-dimensional space, *n* being the number of *other* odors in the set.[77] Such methods reveal "archetypal" odors (Jaubert et al., 1987) that seem to function like primaries (Castro et al., 2013).[78] One instantiation of this odor space, "popcorn," represents a large cluster of similar odors. Still, some

---

[75] *These are not counted in the previously mentioned 400; there are another roughly 300 of these non-working odor receptor genes in humans (Mombaerts, 1999; Ressler, Sullivan & Buck, 1994).*

[76] *The dolphin Stenella coeruleoalba has only non-functional pseudogenes (Freitag et al., 1998).*

[77] *In the Henning prism, an odor is placed on the surface of the prism, and so it is in either a 3- or 4-dimensional odor space, somewhere in between the points, and depending on which odor it is closest to. In an n-dimensional space, the odor is placed in between any number of odors.*

[78] *The authors state that their discovered "primaries" function more as an approximation than a fixed structure. That being said, they are listed as such: fragrant, woody/resinous, fruity (non-citrus), chemical, minty/peppermint, sweet, popcorn, lemon, pungent, decayed. (Forever a glitch in olfaction, notice the need for a disambiguated primary in "Fruity Non-Citrus.")*

notes, such as "anise," remain isolated (Chastrette et al., 1988). In this case, popcorn is "closer" to all of the other odors in odor space, and anise "further away."

Even these types of cross-cancelling methods are unsuccessful in assembling a hierarchical structure to the odor space, because the hierarchies are approximations that change with every instantiation of the odor space, but also because any taxonomy that requires "Other" as a taxon is, for all intents and purposes, useless. This fact leads to the conclusion that "odor space may resemble an *n*-dimensional head of cauliflower more than a simple Euclidean map" (Lawless, 1989: 349). It is premonitory in regards to the potential significance of olfactory knowledge that these processes of information handling yield approximations, not answers. It is by design that these methods are not exactly solvable. A radical redefining of organization itself is required in order for meaning to arise.

To date, there is no clearly delineated categorization of smells, no spreadsheet, only a watercolor painting with vague definition. It is at this point, both within the passage of this text and in the Zeitgeist, that we must begin to take seriously the idea that approximation is a greater means of categorization than analytic deconstruction. Nevertheless, in an attempt to find some sense of stability, a final and much simpler structure of the aromaplex is conveyed.

All smells are composed of carbon and hydrogen atoms. This is a given, because this is what makes them organic, and we can smell only organic molecules.[79] In addition to these two elements, smells can have either oxygen, nitrogen, or sulfur added, and usually only one of these. "Sulfur" smells are exactly what you might think they are – rotten eggs. "Nitrogen" smells are like rotten fish. Smells that have oxygen, or a combination of the others, or even ones that have only hydrogen and carbon, make up the rest of the aromatic volatile organic compounds. Adding oxygen and nitrogen makes something that smells like freshly baked bread (ketones) and that seems to have nothing to do with rotten fish.[80] Beyond the "sulfur" smells (thiols) and the "nitrogen" smells (amines), the unpredictability sets in again. This base-set, however, does offer some understanding of the greater olfactive system: Bad smells tend to contain either sulfur or nitrogen. This is not a hard-and-fast rule. It is, however, at least a basic tendency that can begin to inform further efforts at classifying everything that smells.

The Art

---

[79] *There are exceptions to this general rule, however. Ammonia smells, yet it is not an organic compound.*

[80] *It is possible that one might travel via Nitrogen from fresh bread to slightly spoiled mushrooms to rotten fish, but it is not probable.*

For the majority of history during which the language of smell was being developed, perfumery was both the dominant player and the most fastidious "classifier" available. It is interesting to note that in spite of the significance of hedonic value in the overall categorization of smells, perfumists actually *resist* the hedonic dimension (Roudnitska's unpublished notes, via Holley, 2002: 18). Experiments support Roudnitska's statement – wine tasters show "thinking" activity when tasting wine, whereas most people show more activity in their emotional centers (Castriota-Scanderbeg et al., 2005). However, these professionals do use other binary separations. These separations include the Plant-Animal classes, later collapsed by synthetics and rearranged as Organic-Synthetic. Masculine-Feminine is a separation that relates more to fragrances as opposed to individual molecules, but it still counts as a type of classification. The most recent bifurcation has been that of the Classic-Modern. Again, these terms are used primarily to classify the (composite) fragrances themselves. Nevertheless, it is due to the advent of specific chemicals such as Calone, mentioned previously, that this organizational shift is required.

Alongside these simple two-sided groups, the art of perfumery uses the indispensable concept of accords, which is a nightmare for the methodically minded category maker. Accords can be composed of as little as two notes, which can be added together in different ratios to create different accords that are *distinct in character from their component parts* (Calkin & Jellinek, 1994). For example, combining eugenol and benzyl salicylate in 9:1, 1:1, or 1:9 ratios not only will yield different versions of the combined chemicals, but each version would have its unique scent.[81] From an analytical view, accords are too polycategorical to be rigidly structured. They can be based on places (Oriental), processes (Leather),[82] or even – in a prestidigital act of autopoiesis – perfumes (the classes Chypre and Fougère are eponymously named, as they were first hugely popular perfumes). For the purpose of strict classification, accords are marginally effective at best. For the perfumer, that artist of the most mysterious of senses, they work just fine.

The name "accord" has an interesting etymology in light of modern practices of perfumery. It was originally the task of the perfumer to "bring into harmony" all of the naturally occurring aroma compounds present in a

---

[81] *This example is for explanatory purposes only. That particular combination is a popular two-chemical accord, but the actual resulting aroma-descriptions will not be described in this text, beyond the hypothetical example given.*

[82] *The leather smell is not from the leather itself, but from aromatic materials impregnated into it to cover-up the unpleasant smell of the materials used to treat the leather. An immediate experiment regarding the irreducible odors that subsequently make up the Leather accord, or any accord, would be to visit the amateur perfumer website basenotes.net, and search the forum with a query such as this: "what does _____ smell like." As a fertile tree upon pruning grows new shoots, the resulting comments add more descriptive terms; narrowing-it-down is not the point when it comes to accords.*

mixture. When mixtures contained various essential oils, each one bearing its own subtle recipe, then managing the resulting cacophony was no simple feat. With the appearance of synthetic compounds that could be used separately from their naturally occurring counterparts, the perfumer became tasked with "decorating the monolithic accord" at the center of the fragrance (Calkin & Jellinek, 1994: 139). The craft became less about harmonizing quasi-unintentional by-products and more about rounding out the unnaturally stark odors, reassigning the by-products at will, from the ground up, and via irreducible components.

Nonetheless, keeping with the musicological theme, accords are made of notes. These "notes" can take any categorical form. For example, "civet" (an animal), "eugenol" (a monomolecule), and "clove" (a plant part, the main constituent of which is eugenol) notes are used indiscriminately of their categorical status. In some cases, the line between notes and accords is blurry. Citrus can be an accord, and yet by way of having such a prominent monomolecular representative (limonene), it can also be classified as a note. Notes are used to address one thing that accords do not; specifically, the temporal aspect – the duration of a smell. As a fragrance dissipates, it will go through phases of its odor profile, starting with the top notes, like citrus; then to the body of the fragrance, perhaps a delicate floral, or something a bit more spicy; finally, the region of the odor profile that endures for even days after, still remaining on a holiday sweater you wore while hugging your aunt – that civet, musk, deep, base note. These notes are also termed high, middle, and low notes to communicate their temporal behavior. Most smells fall into only one of these temporal or dissipative categories, because it is one of the more straightforward of characteristics: Larger molecules last longer. Civetone is larger than limonene; one is a bass, the other a violin.

Putting the cap on the bottle of the perfumer's taxonomies, Michael Edwards, a fragrance consultant in the 1980s, penned the "Fragrance Bible." Edwards posited that fragrance should go the way of wine, where it was organized according to cultivar. In the fragrance version, scents were not to be reduced to their botanical origin, but to a kind of "circle of fifths" in music theory, using the intermingling accords rather than individual odors. His *Fragrances of the World* guidebook (1984) collapses the classes given heretofore, such as Masculine-Feminine,[83] but more importantly the Classic-Modern distinction. The impetus for the catalog from an historical-fragrance perspective was to adapt the classic fragrance categories to the new world of synthetics. In reality, however, the catalog was meant to support retailers in their attempt to provide their customers with similar fragrances to the ones they already knew about.[84] Customers have a

---

[83]*Masculine and Feminine categories were not combined until a later edition.*

[84]*It is in line with the general timeline of this larger exposition that customers only know fragrance based on other fragrances as opposed to a list of smells to initially reference other fragrances. Smell does not lend itself well to reductive analysis, gravitating instead towards a*

different language when it comes to smells, so the Edwards' catalog mediates this difference, enabling the Industry and the People to communicate, for the sake of commerce.

The People

In this compartment of olfactory classification, smell is organized by the human organism at large, the flesh of the anthroposphere, if you will. Whereas scientific taxonomies and fragrance industry inventories are themselves products of human endeavor – created for humans by humans – the (unconscious) categorization of odors by the general population is a separate system altogether.

A few studies have been undertaken in an effort to correct the data in the smell literature regarding cultural tendencies and biases.[85] Smell is learned, and so different groups of people respond to odors differently. Any investigation into odor-preference or odor-identification requires knowledge of how these tendencies and biases can rearrange the data. In conducting these studies on cultural preferences, a general classificatory model was exposed.

The first division is predictable at this point in the text. It is an hedonic dyad of good and bad smells. All good smells are either aromas or fragrances, and they are identified quite easily: Aromas are edible, fragrances are not. All bad smells are then subsequently categorized by default and are called "odors." This paradigm is not a new idea on the course of this text. It is reiterated, however, in light of its purpose to the organism. It shows the importance of edibility in olfaction. The fragrance industry, though it diffuses many aromatic arrangements upon society, and though it intermingles with the flavor industry, has its history in toxic concoctions that were nowhere near edible. The pursuits of science, and of chemistry in particular, were not motivated by gustatory desires, and they delivered little to those areas of human experience.[86] For the general population, taste and smell remain intertwined. Thus, half of its system is overshadowed by its relative-sense. Going further, this hedonic dyad of good-to-bad smells is a reminder of the permanent and most primary uses of olfaction: The good-bad dyad translates to the mobile organism as approach-avoid, or ingest-disgust in the case of food.

This observation leads directly to the next configuration of olfaction in the human organism. Firstly, olfaction detects nourishment by discriminating

---

unitary point of identity regardless of its number of associated parts.

[85] *Although the National Geographic Smell Survey is the largest study on this subject, previous work set the basic categories. See: Ayabe-Kanamura et al., 1998; Schleidt et al., 1988; and Seo et al., 2011.*

[86] *With the advent of industrialization and the subsequent development of artificial food products, a vast team of disciplines and entities is devoting resources to the greater understanding of flavor and gustation.*

between nutrients and toxins. Secondly, it activates motility either towards or away from other things in the environment. Finally, it identifies the more complex relationships of kin, mate, and foe. These relationships can be broadly rewritten as Food, Environment, and Humans.[87] Further studies extended subdivisions of this system (*Ayabe-Kanamura et al., 1998; Schleidt et al., 1988; Seo et al., 2011)*, but the primaries still seem to hold. Smells of "nature," "traffic," and "buildings" all fall under Environment. "Cigarettes," "perfume," and "daily products" could rest between Human and Environment. "Waste" is such a broad category that it traverses the entire triad. At what point does food become waste in one's environment? When it slowly rots? When it is digested and then belched? Or, when it is finally excreted? As stated previously, bad smells populate the entire nether-region of the smell taxonomy, and they resist fine delineation (as they resist contemplation, favoring the "speechless" automatic response system instead). When the word for the smell of feces came up as suspiciously vacant from many cross-cultural smell surveys, it was suggested that perhaps it was simply too much of a taboo to talk about. Good suggestion, but maybe not all smells were meant to be named. After all, when we smell some things, our face says it all.

Disassembling the Phantom Lexicon

Herein, the presented methods of classifying smells do not make a clean separation of the full expanse of available scents. The miasma of smell's lexicon can be rearranged according to various organizing principles, each of which provides only part of the true identity of any one smell. Shifting sands such as these cannot possibly serve as a solid foundation. The resulting bunch of tangled knots that are the odors and their webs of meaning reveal the shakiness upon which we build our structures of olfactory knowledge.

In the Bay of Bengal live the Ongee people, a "nose-wise" society who treat olfaction with as much importance as a Westerner does vision. Once, when asked by a scientist for help in making a map of their land, an Ongee man responded: "All the places in space are constantly changing. The creek is never the same; …. Your map tells lies. Places change. Does your map say that?" (Pandya, 1991). Smell is like this. Whether through the meanderings of history, or the chimerical configurations of postmodernity, smell is always changing. The Ongee are right: There are no maps, no categories, and no lexicon to convey this reality.

---

[87]*This tripartite structure of odor taxonomy was presented by Stevenson (2010). He did not use the terms "Food, Environment, Humans," but instead determined the* function *of olfaction in humans as giving rise to ingestion behavior, avoidance of environmental hazards, and social communication.*

The classification of smells has now been folded up, as an origami swan. There is no inherent classifiability in the aromarena. It is organizable to an extent, and depending on its purpose, such designs can be beneficial. Chemists, fragrance specialists, and consumers all benefit from these methods of categorizing smells. The categorization of smell for its own sake, however, is evasive and ethereal. It is like the scent of violet, which intermittently deactivates its own sensor. It is like the adaptation of the olfactory sense, where the more one tries to *know*, the further away the answer appears.

Smell relies so much on our bodies, our memories, and our cultural upbringing that it remains encrypted by a vast network of approximate values, both mildly consistent and highly subjective. Regarding the study herein, the effort is directed not only towards explaining this network. It is also targeted on its points of ambiguity. Standing on the transom of the "gateway to the mind," making sense out of smell reveals something about how we confuse ourselves, and how knowledge is created.

Part Three – CONFUSION

What does it mean to be human? Psychologist Roger Sperry asked this question on a precipice of brain science, one situated between the culmination of behaviorist theory and the dawn of computer-assisted intelligence (1966). Perplexed by what this "multidimensional, intracranial vortex draws into itself," he expressed his concern in recalibrating the way in which we measure ourselves. Today, we sit on a different precipice, one in which the mind, especially the very human part of it, is a computing machine, and the human a potentially replaceable entity.

As theory becomes reality by way of reproduced experiments, and as the scientific literature begins its onward march into the realm of the public, ordinary people are left with confusion over how to integrate these new concepts into their existing worldview. To this point, smell has been presented as a model case of "dirty data." This is an information system we use every day. Nevertheless, it continues to elude our current methods of extracting knowledge from information, serving instead to further confuse us. From this point on, however, the inherent sense of smell is used as a paradigm through which we can view such things as artificial intelligence, multidimensionality, and quantum theory. All of these subjects are on the threshold of public discourse, and they require us to reimagine our "humanist values," as Sperry called them (1966).

How can we explain the inner workings of a machine that teaches itself? How can we explain a world of multiple dimensions, beyond the three we are so familiar with here on Earth and in our bodies? What about the parallel universes of quantum mechanics? These seemingly distant topics of contemporary discourse may benefit from a healthy dose of exogenous material from the olfactory domain. Perhaps to "think like a nose" can offer some advance in these endeavors.

Artificially intelligent systems are now so ubiquitous that we are refining their purpose to be not only intelligent but autonomous – a machine that cannot only think for itself but can generate its own, novel algorithms for thinking. The olfactory system is a unique, intuitive model for a machine that teaches itself, and it shares many characteristics with neural networks and the unsupervised learning algorithms used in artificially intelligent object recognition. This is perhaps the most obvious example of olfaction wafting into another discipline, and this is not the first text to cross the subjects. A technical book published by the Massachusetts Institute of Technology in 1991, titled *Olfaction: A Model System for Computational Neuroscience*, focuses on the neural networks of the nose-brain.

In the Information Age, space is no longer limited to the low-dimensional concept of body space. "Body space" is the number of directions a body can move at once, which is only three. The mind, however, can move in many directions at the same time; hence the distinction between three-dimensional body space and multidimensional information space. High-dimensional data sets – that is, Big Data – are the cornerstone of the Information Age, and they present a major challenge to navigate. The hypothetical smell lexicon, by contrast, is a multidimensional information space that we navigate every day. Looking back on the origin of navigation-as-chemosensation can make us more at home here. Similarly, considering the veritably infinite dimensions of olfactory space can help us to expand our concept of spatialization and to think high-dimensional thoughts.

Although quantum theory has become a very real thing (take macromolecular teleportation, for example), the everyday person and scientists alike continue to call it irrational. There are phenomena in olfaction that exhibit behavior that could metaphorically resemble quantum superposition; that is, the theory that a particle can be in more than one place at the moment before it collapses into existence. Instead of a particle, here we look at the hedonic quality of a molecule and whether that molecule is sensed as a "good smell" or a "bad smell." Although this hedonic dyad is usually the easiest way to begin a categorization of smells, there are some molecules that resist. Thus, after multiple experiments, we cannot predict how a person will report the odorant. Consequently, the molecule cannot be fixed onto the chart. Its box cannot be checked, but instead the checkmark hovers above both boxes. In our attempts to reconcile such impossible worlds, such as the one created by quantum mechanics, we can imagine our collective nose, observe its behavior, and perhaps learn something from it.

Today, as self-generating intelligence machines emulate splintered modules of our brain activity, as the need to manipulate virtual information-space competes with that of physical body-space, and as the very standards informing our knowledge of the universe impress upon us their undeniable-albeit-unbelievable existence, it is time again to check in: What does it mean to be human? Let us pretend that to be human is to smell…

Olfactory Space and n-Dimensionality

Big Data is not about lots of data, but about lots of dimensions to the data. It is hard for people to understand because we have a way of looking at the world using the three dimensions through which our bodies move. There are other kinds of space, however, like the kind where only information lives, and no longer our bodies. This is quite confusing because the contemplative apparatus we use to think in the first place, and to make analogies, is based on the conventional notion of three-dimensional body-space.

Ever the objective of this text, disambiguation must be employed forthright. In order to place the discussion inside information-space and to contend with its multidimensionality, it first must be removed from the more conventional notion of space, that being the empty void that surrounds us and its three-coordinate structure. If we were to walk across a room, then the space we are "moving through" is herein referred to as body-space. Significantly, it is defined in disambiguation of the more comprehensive notion of space from which it is derived, that being the information space of the sensing organism. Body space limits our ability to conceptualize high-dimensional data and the information space in which it exists.

This three-dimensional body space has not always been the default model for navigating our environment. It is the result of, amongst other things, a process of sensory evolution. One of the most distinguished aspects of this evolution was the shift from the proximal, or chemical sensory systems to the distal senses of vision and audition (and was the first distinction to be made, in Chapter 2, when comparing smell to the other senses). For this reason the chemical sense of olfaction provides a unique vantage point from which to consider the origins of space and dimensionality, in the way they existed to an organism that has no sense of distance. The first organisms lived in a world of contact. All of the space beyond their sensory frontier was essentially unknown, because chemosensation requires contact. What is space to such an organism, and what are the dimensions along which it is configured?

Let us begin removing body space from the conceptualization of *n*-dimensional information space using as a point of departure one of the most peculiar texts in olfactory literature. Coauthored by a philosophical neurophysiologist and a comparative literature scholar, the essay "Osmetic Ontogenesis" presents the olfactory system of the closest surviving relative of the ancestor of all vertebrates, the tiger salamander, as a model for the "intimate affiliation between ontogeny and epistemology" (Hosek & Freeman, 2001). In the text, the salamander *becomes* through the dynamic process of sensing chemical information in its environment, perceiving it, executing a decision, actuating motility, and recording the results, which are then used in future acts of perception and decision making, and so on.

Instead of the ontogeny of an existentialist salamander, the following passages will focus on the ontogeny of the conventional notion of three-dimensional body space from the point of view of an evolving organism moving its way up the Tree of Life. From the first eukaryote to the modern human, this organism receives information from its environment and records it as a state space from which it compares subsequent states. These states are the configurations of its information space. The organism uses the differences between states to inform decisions of movement. It is upon this substrate of all possible state spaces that the three-coordinate system of body-space is formed, and it is from initially "moving through" state spaces that the organism eventually recognizes the physical movements of its body through body-space. The objective triangulation of the organism is built on top of the multidimensional information-space, and not the reverse. So let us rewind the tape and observe this chemical epic.

The Odyssey of Body Space

To smell is endemic to the conceptualization of space. The olfactory system encodes patterns of chemical features in the surrounding world into complete body-states of the sensing organism. It then uses that information to react accordingly in all future encounters with the same pattern. Our notion of a Euclidian map is a new concept for an organism that functions in this way. This is first a world of objectless space. The surface of the organism – or more specifically the surface of the sensory apparatus – is the only "place." Distal projection only later creates a virtual distance, a space in-between the organism and other things. It is easy to forget that the outside is always in contact with us. We are perpetually recording this information as well as its effects on us. We think we move through space, and we do. However, the thing that controls us is flickering through iterations of limbic-states, searching for the most apt response to the hedonic gradient. Somehow, we manage to avoid entropic disintegration, one informed actuated potential after another.

There is a eukaryote, swimming in a plasma of organic molecules. The eukaryote is composed of these very same molecules, albeit in a different state of entropy. Flickering about, this organism spreads its random walk throughout its environment. It is literally swimming in a sea of instigatory others, those being the ingredients of the primordial soup.

The eukaryote reacts to the chemical information it encounters in at least two ways – positive or negative. Humans are no different, swimming in a soup of information. Cognition aside (or does the eukaryote think?), we navigate even the complexities of our world in this most primary way. Every piece of information we receive is placed on this hedonic gradient and weighed in light of all the rest until the moment we initiate an action.

The eukaryote does not move of its own accord but, instead, in response to the things outside it. It needs these things in order to move. Ultimately, it does not move itself; they move it. For the chemosensing organism, space is not a void to be traversed. Space is a distribution of the potentiality for movement. The distance between things is not important: There is no distance, only contact.

Chemosensation requires contact between the physical molecules. There is no space inherent in this, not in the way we would understand space. Unless it is in-contact, it does not exist. There is always something outside, and all of it potentiates movement. "Where" is not a concept in this world; there are only degrees of activation and inhibition.

The eukaryote, through its random meanderings, becomes more entangled with the things that surround it. It develops a memory of its interactions, of these entanglements. The memory has a structure. It is a map, but not of the environment. Rather, it is a map of the interactions and their relations to one another. The topography of this map, then, is not one of the environment, but of the weights, the gradients of valence initiated by chemical contact.

It is more difficult than this to understand. There are no single units, or markers, on this map. Every molecule in contact, in every instant, is being measured at once. There is no memory of place, only of the recorded measurements of an instant – all molecules present and their potentials for movement. These instants are plotted together to create the "space" of the chemosensing organism. There is no movement through space. Rather, movement occurs from instant to instant – from one state of affairs to the next. Distance, in this world, is a measure of the difference between these states.

This wandering eukaryotic body exists at a time prior to space as we know it. The maps are of reactions within the organism. They are not maps of the outside world. That which creates space exists at the interface between the inside and the outside: These are the sensory receptors and their ensuing reactions. For the locus of contact, however, that being the organism, the map does not extend beyond this liminal interface. The sensory apparatus is not yet capable of such indirect entanglement. Rather, it still requires contact. There is no distal perception, and therefore no space.

As this organism develops, however, so does its inner landscape. On the brink of three-dimensional body-space the memory expands, tightening its omnitethered fibers. Whole instants of multimolecular contact are no longer matched against one another in sequence. The instigating plasma, in each instant, is separated into parts, and these parts are recognized, or "remembered" upon subsequent instants. Thus, a whole system of comparisons is being formulated by the innerworkings of the organism's memory. A greater understanding is being derived; the inside and the outside become less equally distributed, and the body expands as well. The memory is not a brain, as we would now understand it. The memory is the body. In a backward sense, cephalization does not appear and develop in the body of the organism. Just the opposite: The body grows out of it. The body is, today, the part of the environment that yesterday the mind decided to take into itself.

As organism and environment become each other, the ability of the organism to predict or at least maintain virtual instantiations of its world improves. The organism that is essentially made of its own surroundings has already "been there, done that." It chose what to take and what to leave behind, and it maintains a record of every transaction. The organism, of course, has not yet developed the ability to manipulate this record. At this stage, the writing is in stone, encoded into the body. It cannot be changed. The plasticity of a virtual body has yet to appear.

Looking further into the future of this organism, it creates within it a map so complex, so impeccable in its detail, that many of the once exogenous components of its environs eventually become part of it. Or, rather, together they create something else – made of both, yet apart from both. This is the pattern of development: The overall landscape of the distribution of molecules changes, and the organism, though still ultimately made of the same plasma in which it operates, favors the integration of certain molecules around it, leaving the outside to be of an increasingly different organic signature other than its own. Through their interactions, they change each other. The organism, through successive interactions with its environs, over the course of successive generations, becomes so "informed" that the two become incrementally interchangeable with each other, via the interfacing membrane. Patterns of molecules create corresponding receptors, which in turn change the way the organism interacts with those molecular patterns. The plasma, once fluid, has now fractured into solid identities to be encountered.

We can barely see it now, the future of our allegorical organism, because it is almost under our nose. The organism's ability to discriminate amongst the myriad outside encounters is refined. Its memory is deeper and denser. With it, the plotting of configuration states becomes so interconnected that a new map precipitates out of the recognized patterns. Now moving through the plasma, the organism is armed with a device for recognizing entire ensembles of organic information against the background of molecular noise. These objects are organized according to a new kind of distance, and they are redistributed in a new kind of space. This is a space where universality is required to reconcile the simultaneous configurations of more than one object in relation to the organism. This is a shared space, its common ground demarked by the relations of the objects to one another.

This is body-space, or "three-dimensionality." It requires three objects altogether – one subject, or recording device, and two objects to be recorded relative to each other and each to the recording subject. This resulting information triangle begets "location." Space is then no longer a distribution of the potentiality for movement, as instigated by the sum of direct, outside encounters. It is now a distribution of all potential encounters. These potentials are assigned equal value; in other words, they are distributed equally, as are the regular coordinates of the three-dimensional Euclidian system. The organism can then move relative to this grid and the positions of the objects within it, and not to the objects alone. The preexisting maps of the relationship between the organism and the molecules that surround it are registered against a structure holding all possible interactions simultaneously. This latticed cube becomes a new form of representation. The molecules – and their effects – are no longer the primary elements of the map. They can now exist separately altogether, and they are referenced by way of a different organizational structure of three-dimensional coordinates.

The eukaryotic Odysseus finally acquires the distal senses of vision and audition, expanding its perceptive field to a world never before imagined. Highly processed signals from photo- and mechano-receptive sensors enable the body to interact with its environment in ways that will go on to make chemical communication seem like a primitive means of survival. In turn, the cortical areas of these distal senses, unlike olfaction, will find their way out of the limbic system and will grow to enormous size. Eventually, once every instrument in the orchestra is tuned together, the symphony of conscious self-identity will begin.

To perceive space from the point of view of this eukaryotic organism, as a distribution for the potentiality for movement, is to peer deeper into the true nature of space and to disentangle the meaning of dimensionality as confined to the three dimensions of the movement of the body. Intimately integrated with the limbic system – mission control for body movement and spatialization – olfactory information activates movement and is thus endemic to the conceptualization of space. So automated is our response to olfactory information, or most sensory information for that matter, that we forget three-dimensionality is only the topmost layer of an incredibly high-dimensional map of information-space. In the same way a two-dimensional map helps us to locate things in three-dimensional physical space, body-space is a dimension-reducing shortcut for the tremendously complex reality of multimodal sensory information. The transition from the proximal to the distal occurred quite some time ago, so much so that we may have forgotten what is it was like to be the eukaryote.

Navigating the n-Dimensional Depths of Olfactory Data

There is another kind of space other than the one our bodies inhabit. The other kind of space is the one inhabited by the mind, or by any decision-making or information-processing system, whether it controls a body or not. It could be the space where ideas are located, or a lexicon, or simply data points. Whatever their nature, they are located in this space in relation to each other.

Most people know how a phone book works, so as the first alternative to three-dimensional body-space, let us begin with the low-dimensional "telephone book space." A telephone book contains lots of data, but in only one dimension. The phone numbers are the single attribute given to each entity in the database. As a network graph, this looks like a zipper, with each person connected to only one number, and vice versa. Add as much data as you like; a phone book will never become Big Data, because the information it yields is only one-dimensional. Instead, Big Data is more like a telephone book that lists, in addition to names and numbers, the location, gender, ethnicity, fingerprint, faceprint, bloodtype, genotype, level of education, political affiliation, purchasing history, search history, every phone call ever made and to whom and for how long, and so on, and in realtime. This is a tremendous body of data. What makes it Big, however, are the dimensions. It has *n*-dimensions, where *n* equals the number of attributes added.

Eventually, when there are as many attributes to the entities as there are entities, it is as if the attributes become entities themselves. At this point they can be referred to simply as data points. As a network graph, this looks not like a zipper but a ratty bird's nest of entity/attributes connected to one another. Together, they form a particular configuration, or state of the information space. Every data point is assigned its own dimension, so that the resulting body of information takes form in the way a cube precipitates out of a set of three-dimensional data. It is not the data points themselves that describe the space, but the relationships among them. The information yielded by this space is not the locations of objects but the relationships, or the configurations, of the *n*-dimensional space itself.

Our mind does not function in body space alone. Consequently, neither does our system of olfaction, and especially the semantic features of that system, those being the "language of smell." This time, the example comes not from a phone book, but from one of the biggest of the Big Data, that of olfactory space.

Prefacing this book is a photograph of a young person smelling an odor sample in a lab. He has been asked to identify what he smells. As he thinks, where does his mind go? Searching through the autobiographical associative memory network of people, places and things, across sensory modalities and against the lexicon of available descriptors, this is the act of odor-object recognition: iterating configurations of an information space determined by olfactory-related data, and matching it against incoming stimuli. What is this thing I smell?

The space in which one is searching in this instance is olfactory space. It can also be understood as a map of the associative memory network, which includes not only the semantic memory but the episodic memory as well. It is organized, yet shrouded in complexity, and slippery in the cold hand of knowledge. In its totality, this olfactory space contains all of an individual's memories, down to the level of the most pristine physiological detail. The information in this space is the raw data of the entire life of one person. The sheer size of this space is beyond mortal comprehension, and its configurations induce contemplative paralysis. And, this is the olfactory space of just one person. The "universal" or shared olfactory space is the multiplication of even this. To the casual reader, this should sound a lot like Big Data, that disembodied monster that has come to life through the hypermutated exocortex that we made for it. We must not forget, however, that our mind as a whole was not always limited to such low-dimensional data. In fact, the primitive parts of us are inherently aloft in a higher-dimensional world.

Olfactory sensation exists as a constellation of associative memories when it is still inside the individual. It becomes a word when it is verbalized, or uttered forth into the shared world of language and objective knowledge. But smells, unlike words, are not discrete things. They are not like objects in space. As information, smells are more like a space unto themselves, and less like the objects within it. They are a snapshot of a complex interaction, a molecular signal and its stimulating effect on the body of an organism, by way of memory.

The points in olfactory space are components of a particular smell. They may be the different chemicals that make up the total profile of the smell, or they may be memory associations previously attached to the smell by the subject during ontogeny. In either case, however, identifying the components does not tell us what the smell is. The smell is the configuration of the components signaling variations in their relation to one another, both the chemical mixtures themselves and the subjective meaning we associate with them. To add or subtract the amount of only one chemical will change the smell, but not because of the direct effects of that chemical on the receptors. The change comes from the reconfiguration of the entire smell map as it exists in the subject's mind.

On either side there are odors and there are odorants, to use the dichotomy presented in olfactory science; one is an object and the other a percept. A characteristic combination of molecules will initiate a corresponding perception of an odorant. But the smell is not in the molecules alone. Instead, they need the subjectively smelling organism to be complete. In the case of olfaction, the perception of the stimulus is critically dependent on the memory network of the individual: Every odorant smells different to every person.

Objectively, any particularly identifiable odor is instantiated as its own information space of infinite dimension. The most recent count rests at one trillion possible odor molecules. This is not meant to be a real account, because every possible combination of these trillion molecules creates another set altogether. The count of identifiable odors is limited only by the words we can generate in an effort to identify them. Which would be exhausted first – a list of odors, or the words used to make such a list?

We attempt to build the phantom lexicon anyway, by eliciting verbal responses to presented odors. The names or words or verbal utterances or even nonresponses are all made contiguous to one another in a complex lexicon for identifying odors, which is itself laid atop the associated episodic memory network of an individual. Attempts are then made to classify and organize these subjectively described percepts according to primary odor classes. This effort imposes a hierarchical topography over the vast ocean of aromas, situating "like" odors near one another. Floral seems to be a primary, a peak of the topography. It contains all flower-smelling things in its valleys. But, at what point does Orange stop being Citrus and start being Floral? And if Woody is also a category that contains Orange, then where is it located on this map? And what about Clean, or Sour, or Summer, or Breakfast?

Peach is somewhere between Fruity and Butter, and Sour is between Fatty and Fruity (olfactively speaking). It is as if the points exist only for things to be in-between and not to be used as unique classifiers in themselves. Olfactory space is an *n*-dimensional map of all of the possible odor chemicals (which is impossible).The resulting information – that is, the smell – comprises the relationships among all of the points on the map. It is not distance; it is configuration, arrangement. A smell is thus located somewhere between every other point in this olfactory space, a single value obfuscated by its *n*-dimensionality.

The episodic memory, encoded by the limbic system into the body, is at the interface of this olfactive phenomenon (odor-odorant, object-percept). Odors are unlimited by the capacity for organic molecules to combine and recombine. Odorants as semantic instantiation are unlimited by the same combination and recombination of language. The limbic corpus of the individual marks the limit of this overall system. Perhaps "limited" is not the appropriate language to introduce such an entity. The episodic memory (by way of the hippocampus) holds a simulable record of all of the points in Euclidian space where a body was ever located, made contiguous not only by their physical proximity to one another, but also by various co-occurring states – temporal, emotional, gustatory, etc. Within the confines of this expansive limbic reticulum, an individual travels through a simulation of the shared, objective, Euclidian space along distances determined by autobiographical memory-space. The Euclidian space, the hometown, the body-map is folded-up and collapsed on itself, then reconfigured again, one aromatic compound after another. In recollection, sans vision, sans audition, we are yet free to travel the *n*-dimensional space of olfaction.

Investigating the Artificial Unconscious

Instructions: Find a tangle of knots. It should not be too hard – earbud wires seem to have a thing for breaking down into the tangles of entropic disorder. Next, untangle the knots – but not with such direct intent. Do not find one end of the wire and needle it in reverse, or spot a cluster and pull at it. Instead, take that messy knot of wires, and stick your fingers into it, all of them, right into the middle of it. Now begin to move your hands in and out, while wiggling your fingers. As if the knot of wires is a blob of matter and you are massaging it, let your fingers dance with the knot, grabbing, pulling, looping, crossing. It is not efficient, and it does not take advantage of that human capacity for self-directed behavior, but it works. It is semidirected, partially intentional. If you let your fingers do most of the thinking, then the knot will eventually untie itself. In the process, you will have seen the unconscious mind in action.

This is also the act of odor-object recognition. It is not a record, linear, of every chemical present. The signal-to-noise ratio for chemosensation is too complex. Instead, it is a whirling chaos of switching and regrouping that produces the answer in a way that makes absolutely no logical sense whatsoever. In fact, it is here that the seeds of logic are sown, and so, only *after* the answer has been found, the fruit.

A group of children placed together, but with no exposure to language, will eventually create their own dialect. The structure is there. No instructions are required for creating a language, only the capacity for verbal communication and a whole lot of babble. Deep learning neural networks are a second-generation artificial intelligence that utilizes "hidden layers" of bottom-up information flows. Top-down instructions say "If you see this, then do that, unless it is like this, in which case do that instead." By contrast, the bottom-up script says, "What do you see?" Unsupervised learning, it is formally called in the field of artificial intelligence. Using these "hidden" structures, the functional neural network will tell *you* what features to consider. The network will organize itself, or rather the *data* organizes itself, using the available network.

Artificial intelligence is a matter of information metabolism. Either the system gets nothing in terms of rules, but everything in terms of information, or it gets specific rules that work only with specific forms of information. If it can gain access to infinite information, it can then generate its own rules, for every instance no matter what (in an ideal system). The work then goes into providing information, not to "figuring it out." The deep learning artificial intelligence is a machine, a fulcrum that multiplies force, and it is fed not by way of fossil fuel, but by using a different kind of fossil – our cultural artifacts, be they odysseys of antiquity or pictures of cats.

The olfactory system does not produce cultural artifacts so much as individual ones. All the same, the data sets are big. Trillions upon trillions of molecules, all linked to body-state and immeasurable autobiographical data, matched against a set of words that can take virtually any form the language can provide and that must be subject to the generative properties of language in that a new word can be used that did not previously exist. These are the data sets involved in olfaction. They are big, and they are yet manageable, operable, and capable of producing knowledge about the world around us. The nose knows something we seem to be only now figuring out: With enough data and the right substrate, knowledge creates itself.

Furthermore, the rules that govern this processing do not come predetermined in the human organism. Smells are learned, not given as fact. Olfactory perception generates its own rules based on the data sets given to it. There exists no such thing as a universal response to any particular odor. At times it may seem so – *menthol is cool* – but this is trigeminally determined. *Burning flesh is bad*, but this is culturally determined. In *any* case where it seems that an odor response is preinstalled, it will turn out that either the boundaries of the system are ignored (as in the trigeminal discrepancies), the power of cognitive override is underestimated (as in suggestively induced odor hallucinations), or the experimental group is not large enough (to include cultures both above and below the water table where burning bodies is more practical than burying them, for example). In fact, there are many reasons why a smell may *seem* to have a quality that is universal to all humans. However, because olfaction is so incompletely understood, by professionals and the general public alike, this universality is always an illusion, painted by our lack of knowledge on the subject.

Olfaction utilizes unsupervised learning on neural networks of extremely massive and complex data sets to encode and subsequently recognize odor objects. Nevermind reason – to assign such a high degree of intelligence to such a primitive sensory system had to await the development of its artificial analog.

Quantum Hedonics

Perhaps it was excusable 50 years ago to say that "nobody understands quantum physics." We have come a long way since then, however. Who would have thought that the quirks of the quantum world would be found in the photosynthesis of plants? But alas, when an electron from the sun enters a leaf, it samples the routes available, smearing itself across all of the paths in a state of quantum superposition, searching for the best route for delivering its solar energy to the photosynthesis parts of the plant (Hildner, 2013). This entire process lasts on the order of one-quadrillionth of a second. Little pieces of the sun are performing quantum computer search algorithms on the leaves of plants.

Time has run out, and the time is now to overcome the discrepancies of quantum theory as unnatural, nonsensical, and counterintuitive. It is all of these things, of course; but we no longer have any choice in the matter. Truth does not ask permission. The Earth is *not* flat, the planets are *not* pushed by heavenly creatures, and air is *not* empty. Rational thought has led us beyond these frontiers, and rational thought asserts that "a thing *cannot* be in two places at once."

Of all things, there is a peculiar olfactory phenomenon that causes psychological effects very similar to this aspect of quantum mechanics. Isovaleric acid – it is the smell of sweaty gym socks, vomit, and Parmesan cheese. It is the smell of dirty baby diapers, and Camembert. It is also one of the most exceptional cases of "quantum" hedonics in all of olfaction. If a test subject or, rather, 100 subjects are presented with a vial of isovaleric acid, the statistical distribution of their hedonic responses will read as such: Half will call it scrumptious, and the other half will gag. There are no in-betweens, and there are no biases. How, then, can this smell be in two places at once? How can it be categorized as both good *and* bad? Technically, if we were to attempt to predict the response, there would be a 50/50 chance of either answer. Isovaleric acid therefore does not exist on the hedonic dyad of olfaction because it cannot be determined with any certainty as to which position it will inhabit.[88]

Certainty, however, is a product of Classical measurement. Approximation is more appropriate. For example, we know that isovaleric acid will not fall in the middle of the hedonic spectrum. It is almost never reported as

---

[88] *This is an extremely rare condition as far as smells go. Most smells tend to be consistent in their hedonic effects (within a "culturally contingent" test group). Their statistical distribution is rarely spread out, and when it is, there is some explainable, learned reason for it. Things like fish sauce, sauerkraut, and durian fruit probably exhibit the same "quantum" effect (on their corresponding populations). The reason isovaleric acid is so often cited in smell literature is probably because most of the studies are done on a Western-influenced audience that is familiar with aged cheese.*

"neutral." We know that there is a 50/50 chance of it being at one end or the other (delicious-disgusting). This is not certainty, but it works. As far as we know, the smell is in two places at once. This needs explication, however. It is not that the smell *is* in two places at once, but that it *can* be. Quantum theory does not necessarily say things are in two places – that is its quantum state, which eventually collapses. Once a thing becomes such, there can be only one.

Classical causation is under suspicion from yet another angle. Quantum theory posits that despite approximation and imperfect predictability, if we were to limit our observation of the photon to the one place, it would be there; yet if we were to look in the other place, it would be there instead. In fact, if we were to look in only the one place 100 times, it would be there 100 times, because we *make* it there by the act of *looking* for it.

In the same way, if the experimenter emits even the slightest bit of contextual information into the test, the hedonic duality of isovaleric acid will collapse. If it is named one way or the other, if it is presented *after* a good or bad smell, if it is presented in *any way* that disturbs the pure, "quantum" nature of the smell, it collapses.

Another olfactory analogy – white smell is the paramount of olfactory confusion, and so significant a discovery that it warranted a Nobel Prize.[89] Using a protocol set of monomolecular odorants, the scientists adjusted the perceived intensity of each smell to make them all equal, as well as to place the smells equally in the olfactory space of odor similarity. The resulting components, all equal in intensity, all chosen from an equal distribution in olfactory space, were mixed with one another and then tested on their pairwise similarity by sniffing human subjects.

The scientists discovered that the more components that were added to a mixture, the more it smelled like any other mixture, *even though the mixtures had no components in common.* Imagine mixing the representative aromatic molecules of apple, banana, cinnamon, and nutmeg in one flask; and semen, sweat, civet, and excrement in the other. Now imagine that the two mixtures smell the same. This is an extremely simplified and technically incorrect example, but it is the basic idea behind white smell.

It is already well known that even trained professionals cannot discriminate among the components of a mixture of more than a few monomolecules. However, this experiment, by equalizing the intensities and the odor-similarity of each component, goes one step further. It reveals a vast, universal characteristic of all odors. This "characteristic" is a peculiar one,

---

[89] *"White Smell" is credited, among others, to a man named Weiss (German for "white"), from a place called the Weizmann Institute.*

in that it does not reside in any individual odor, but in their combination with other odors. Somewhere between the tenth and the twentieth component being added into the mix, it is as if the individual odors have disappeared completely, into the "white" of olfaction.[90]

In contrast to isovaleric acid, white smell is reported as neither pleasant nor unpleasant, neither disgusting nor delicious. Rather, it is both; it falls right in-between, the definition of ambiguity. If we were to predict the outcome of a subject's hedonic response upon smelling "white," the chances of being correct would no longer be fifty-fifty. Certainty is still evasive. No one can say what the smell actually is. The descriptions given by professional perfumers were highly variable; the smell has a quantum superposition of potential descriptions, if you will. In other words, it does not smell like anything, and yet it smells like everything, at the same time. However, if the experimenter were to tell the perfumer that this white smell has a name, then every time thereafter the perfumer would call it by that name, and the quantum state will again have collapsed.

Whether it is the whereabouts of a photon, or the hedonic response to an olfactory stimulus, there will always be a point beyond which our structure of knowledge on a particular subject can no longer deliver accurate measurements. In a phenomenon where everything is contingent upon everything else, where everything exists as an approximation only and not as a set of discreet components, there can be no objective truth upon which to generate future predictions. There is only a substrate, or a probability distribution upon which a subject must act. The subject and the phenomenon become one by way of interaction, and only as a result of this process does the phenomenon acquire its measurement.

To translate quantum phenomena into a language of olfaction, to toggle one's thoughts between two vast systems, comparing and contrasting, joining and refitting – stranger stories have been told in the struggle for understanding.

---

[90] *"White" is in reference to the behavior of all colors to make white light, as well as all audible tones of equal intensity to make white noise.*

Conclusion: Lingua Anosmia

All of the other senses can be quantified, and hence measured. Smell is wrong in this way. It has no discrete values. Instead, it exists as an approximation among a multitude of vertices, those being, in short order, the veritably infinite set of organic molecules, the infinitely varied set of autobiographical and limbic data, and the infinite generative abilities of language. Together, these properties of the olfactory phenomenon make it the only sense that cannot speak for itself.

By some kind of relative measurement, to smell is the only means for a human mind to make contact with, to simulate, or to *be again* that first organism to *brain* itself out of timespace. That Proustian nostalgia we feel, perhaps, is attached to much more than our own emotional biography. Smell is the animal inside us. But more than that, it is the dawning of a mind. The language of this sense sits forever at the tip of the tongue.

Olfaction is an entirely "unconscious" process. Try as we may, we cannot exercise our olfactive powers at will. We cannot "imagine" smells; they must be exogenously stimulated. We cannot *not* smell, lest we *not* breathe. The more we try to smell something, the less we smell of it, because of its high attenuation rate. To smell is a cognitive asymmetry, taking place on one side of the brain only. Olfaction performs its Herculean feat of pattern recognition and limbic response completely separate from the logic parts of the brain. If the crosstalk between hemispheres is the substrate of our sense of self, then olfaction does not take orders from us. Rather, it is the reverse.

In comparison to the human neocortex, olfaction is considered simple, a primitive affair. The smelling brain is overlooked in the light of such neurological virtuosities as abstract thought and self-awareness. However, there is something to be said for an apparatus that can do so much with so little. At one point in the development of the organism, olfaction was the height of sophistication. It was at this point where the story of cognition began, for the olfactory system is a brain unto itself.

As it turns out, we know less about this nose-brain than we do about the neocortex. The epithelium, the sensory surface upon which aroma compounds are initially received, exhibits no recognizable patterns in relation to specific odors. At the next interchange, the activity observed in the olfactory bulb is almost entirely shrouded in mystery. Let this be a kind of relative indicator: Retinal and cochlear prostheses have already been developed, a feat of reverse-engineering of the nervous system that has decoded the visual and auditory receptors of the brain. In contrast, the translation of smells from the tip of the brain's tendrils in the nose to the percept of a smell in the mind of the subject is incompletely understood. Prosthetic noses will appear in speculative fiction plotlines long before they appear in national headlines.

Beyond this, the later stages of olfactory perception include the limbic system – which is related to the movement of the body – emotion, and spatialization (among other things). We know more about these modules than we do about the initial olfactory interchanges. However, smell is so deeply embedded into the circuitry of the limbic system that the mystery of olfaction reveals a shortcoming in our understanding. We will never fully understand how a mind creates a self, or even how it moves a body through space, without first deciphering the hidden network layers of the olfactory bulb.

Perhaps it should stay hidden. Nonetheless, the odor somehow makes its way to the language centers of the brain, now searching for an appropriate response to communicate the identity of the odor. It is only at this point where olfaction enters the left hemisphere and an explanation of its behavior can come forth (explanations themselves being such a left-brain phenomenon). In eliciting a verbal response to a presented odor, the olfactory system performs a kind of Hamming distance operation of both the episodic and semantic memory data sets, comparatively. The sets are narrowed down until a final word is generated. This word we call knowledge. But is it really?

If knowledge is supposed to tell us which is which, and what is what, then how do we use it to study a thing that is inherently ambiguous? Smell is such a thing. In it, we have an example of an information-processing system that makes its sole purpose to ascertain ambiguous information. Moreover, during the entire process from primitive sensation to cognitive verbalization, it is fuzzy, noisy, and dirty.

The title of this conclusion, *Lingua Anosmia*, is presented in the spirit of the master analogy maker Douglas Hofstadter, author of the metalogical manifesto *Gödel, Escher, Bach: An Eternal Golden Braid* (1979). In searching the hidden mechanisms of the mind for the origins of thought, Hofstadter returned with the infinitely looping Möbius strip as a symbol of the paradox of self-reference.

In the course of this exposition, the "language of smell" has been presented as its own paradox. *Lingua Anosmia* translates as the tongue that cannot smell. When taken literally, it is meaningless. As an analogy, however, it represents an interaction between olfactory phenomena and a hypothetical corpus of the olfactory lexicon as an ideal laboratory for probing our incipient cognition at work.

Stare at a picture for hours, undisturbed but for the fragmentary blips of your lens-moistening eyelids. To smell requires breathing. To smell is indistinguishable from breathing. Not only does it come in whiffs, and is ethereal, but it is always fleeting, both in shared timespace and in the attenuating nose-brain. One cannot grasp an odor by its nature (but for eating it, of course). Viscerally symbiotic, it operates independently of our thinking selves. The sparkling jewels of cognition it ignores. To comprehend, to abstract; to postulate and to imagine; smell is not of these things. It is bound to the body, our skin turned inside-out and smart-side in. It is the first cognition, surely, the primary version. The prototype for the contemplative apparatus to come, it opens the way for the disturbances of the shared medium, in the forms of vision and audition. In human olfaction we are on the precipice of thought, transfixed in the moment between a world bounded by physical reality and the present and one in which we float freely through the timeless, infinitely expansive space of the mind.

Lewis Thomas, etymologist, essayist, physician, and poet, was right when he remarked that smelling a thing is remarkably like the act of thinking itself. To ponder this transformation, the crystallization of pre-thought into thought, of olfactory experience into language, is to think about the most underlying operation of the deepest parts of our mind. To analyze, to separate, to identify, classify, and organize such a phenomenon – such a subjective phenomenon – is to turn the tool of thought back on itself. The practice of experiencing and expressing this phenomenon is our Eternal Golden Braid, and one that every smelling person can identify with and engage with. Now may your thoughts be more about themselves than the things they are about, and may you never smell the same again.

Acknowledgments

It is with great appreciation that I acknowledge my informal editor, Ralph Iansito, my artist, Joe Scordo, and my designer-at-large, Theresa Gjertsen. Without them, this book would be nothing but words on white typing paper. Without my developmental editor, Robert Weiss, this second edition would still be as full of quirks and confusion as the first. I must also thank Anthony C. Mauro for telling me the title of the book. I thank the friends and family who have encouraged me; they have sharpened my focus and humored my enthusiasm. Finally, I thank the one who will go unnamed, the one who taught me the craft of surreal math and the art of abstract food. One does not have to be a genius to understand a genius, only to listen long enough.

Citations

Ackerman D (1990). *A Natural History of the Senses*. New York: Random House.

Aftelier's Natural Fragrance Wheel (2013). http://www.aftelier.com/aftelier-fragrance-wheel.html. Accessed Dec 20 2013.

Alaoui-Ismaïli O, Vernet-Maury E, Dittmar A, Delhomme G, & Chanel J (1997). Odor Hedonics: Connection with Emotional Responses Estimated by Autonomic Parameters. *Chemical Senses* 22: 237-48.

American Society for Testing and Materials, ASTM (1968). *Correlation of Subjective-Objective Methods of the Study of Odors and Taste*. Philadelphia: American Society for Testing and Materials.

Amoore J E (1969). A Plan to Identify Most of the Primary Odors. In: *Olfaction and Taste III*, ed. C Pfaff-Mann, pp. 158–71. New York: Rockefeller University Press.

Arctander S (1969). *Perfume and Flavor Chemicals*. Montclair, NJ: Arctander.

Ayabe-Kanamura S, Schicker I, Laska M, Hudson R, Distel H, Kobayakawa T, & Saito S (1998). Differences in Perception of Everyday Odors: a Japanese-German Cross-cultural Study. *Chemical Senses* 23: 31-38.

Ahn Y-Y, Ahnert S E, Bagrow J P, & Barabási A-L (2011). Flavor Network and the Principles of Food Pairing. *Nature: Scientific Reports* 1: 196.

Basenotes (2014). www.basenotes.net. Accessed July 10 2014.

Berlin B & Kay P (1969). *Basic Color Terms: Their Universality and Evolution*. Berkeley: University of California Press.

Boisson C (1997). La Dénomination des Odeurs: Variations et Régularités Linguistiques. In: *Olfaction: Du Linguistique au Neurone*, ed. D Dubois & A Holley. *Intelligentica* 1(24): 29-49.

Bushdid C, Magnasco M O, Vosshall L B, & Keller A (2014). Humans Can Discriminate More than 1 Trillion Olfactory Stimuli. *Science* 343(6177): 1370-1372.

Cain W S & Johnson F (1978). Lability of Odor Pleasantness: Influence of Mere Exposure. *Perception* 7: 459–465.

Cain W S (1987). Indoor Air as a Source of Annoyance. In: *Environmental Annoyance: Characterization, Measurement and Control*, ed. H S Koelega, pp. 189-200. Amsterdam, The Netherlands: Elsevier.

Calkin R A & Jellinek S (1994). *Perfumery: Practice and Principles*. New York: Wiley.

Castriota-Scanderbeg A, Hagberg G E, Cerasa A, Committeri G, Galati G, Patria F, Pitzalis S, Caltagirone C, & Frackowiak R. (2005). The Appreciation of Wine by Sommeliers: A Functional Magnetic Resonance Study of Sensory Integration. *Neuroimage* 25: 570-78.

Castro J B, Ramanathan A, & Chennubhotla C S (2013). Categorical Dimensions of Human Odor Descriptor Space Revealed by Non-Negative Matrix Factorization. *PLoS One* 8(9): e7328.

Chastrette M, Elmouaffek A, & Zakarya D (1988). A Multidimensional Statistical Study of Similarities between 74 Notes Used in Perfumery. *Chemical Senses* 13: 295-305.

Chrea C, Grandjean D, Delplanque S, Cayeux I, Le Calvé B, Aymard L, Velazco M I, Sander D, & Scherer K R (2009). Mapping the Semantic Space for the Subjective Experience of Emotional Responses to Odors. *Chemical Senses* 34: 49-62.

Corbin A (1986). *The Foul and the Fragrant: Odor and the French Social Imagination*. Cambridge, MA: Harvard University Press.

Damasio A (2010). *Self Comes to Mind: Constructing the Conscious Brain*. New York: Pantheon Books.

Darwin C R (1859). *On the Origin of Species by Means of Natural Selection, or the Preservation of Favoured Races in the Struggle for Life*. London: John Murray.

David S (1997). Expression des Odeurs en Francais: Analyse Lexical et Representation Cognitive. *Intelligectica* 24: 51-83.

Davis J L & Eichenbaum H, eds. (1991). *Olfaction: A Model System for Computational Neuroscience*. Boston: Bradford Books/MIT Press.

Deutscher G (2011). *Through the Language Glass: Why the World Looks Different in Other Languages*. New York: Metropolitan Books.

Dubois D & Rouby C (2002). Names and Categories for Odors: The Veridical Label. In: *Olfaction, Taste, and Cognition*, ed. C Rouby, B Schaal, D Dubois, R Gervais, & A Holley, pp. 47-66. Cambridge, U.K: Cambridge University Press.

Dubois D (2007). From Psychophysics to Semiophysics: Categories as Acts of Meaning, a Case Study from Olfaction and Audition, Back to Colors. In: *Speaking of Colors and Odors, Converging Evidence in Language and Communication Research 8*, ed. M Plumacher & P Holz, pp. 168-184. Amsterdam: John Benjamins.

Duhigg C (2012). How Companies Learn Your Secrets. *New York Times*, Feb 16. http://www.nytimes.com/2012/02/19/magazine/shopping-habits.html. Accessed July 1, 2014.

Durham W H (1991). *Coevolution: Genes, Culture, and Human Diversity*. Stanford, CA: Stanford University Press.

Edwards B (1979). *Drawing on the Right Side of the Brain*. New York: Tarcher/Penguin.

Engen T (1982). *The Perception of Odors*. New York: Academic Press.

Engen T (1987). Remembering Odors and Their Names. *American Scientist* 75: 497-503.

Engen T (1991). *Odor Sensation and Memory*. New York: Praeger.

Gladstone W E (1858). *Studies on Homer and the Homeric Age*. London: Oxford University Press.

Gordin M (2015). *Scientific Babel: How Science Was Done Before and After Global English.* Chicago: University of Chicago Press.

Gottfried J A (2009). Function Follows Form: Ecological Constraints on Odor Codes and Olfactory Percepts. *Current Opinion in Neurobiology* 19: 422–429.

Green D M & Swets J A (1974). *Signal Detection Theory and Psychophysics.* New York: Wiley.

Grossen M (1989). Le Contratimplicite Entre L'expérimentateur et L'enfanten Situation de Test. *Revue Suisse de Psychologie* 48: 179-89.

Haller R, Rummel C, Henneberg S, Pollmer U, & Köster E P (1999). The Influence of Early Experience with Vanillin on Food Preference Later in Life. *Chemical Senses* 24: 465–467.

Han-Seok S, Guarneros M, Hudson R, Distel H, Byung-Chan M, Jin-Kyu K, Croy I, Vodicka J, & Hummel T (2011). Attitudes toward Olfaction: A Cross-regional Study. *Chemical Senses* 36(2): 177-187.

Held R (1989). Perception and its Neuronal Mechanisms. *Cognition* 33: 139-154.

Henning H (1916). *Der Geruch.* Leipzig: Barth.

Hertz R (2007). *The Scent of Desire.* New York: William Morrow.

Hildner R, Brinks D, Nieder J B, Cogdell R J, & van Hulst N F (2013). Quantum Coherent Energy Transfer over Varying Pathways in Single Light-Harvesting Complexes. *Science* 21: 1448-1451.

Hofstadter D (1979). *Gödel, Escher, Bach: An Eternal Golden Braid.* New York: Basic.

Holley A (2002). Cognitive Aspects of Olfaction in Perfumer Practice. In: *Olfaction, Taste, and Cognition*, ed. C Rouby, B Schaal, D Dubois, R Gervais, & A Holley, pp. 16-26. Cambridge, U.K: Cambridge University Press.

Hosek R J & Freeman W J (2001). Osmetic Ontogenesis, or Olfaction Becomes You: The Neurodynamic, Intentional Self and Its Affinities with the Foucaultian/Butlerian Subject. *Configurations* 9: 509–541.

Howes D (2002). Nose-wise: Olfactory Metaphors in Mind. In: *Olfaction, Taste, and Cognition*, ed. C Rouby, B Schaal, D Dubois, R Gervais, & A Holley, pp. 67-81. Cambridge, U.K: Cambridge University Press.

Hudson R & Distel H (2002). The Individuality of Odor Perception. In: *Olfaction, Taste, and Cognition*, ed. C Rouby, B Schaal, D Dubois, R Gervais, & A Holley, pp. 408-420. Cambridge, U.K: Cambridge University Press.

International Union of Pure and Applied Chemistry, IUPAC (2013). http://www.iupac.org/. Accessed Dec 18, 2013.

Jacob S, Zelano B, Hayreh D J S, & McClintock M K (2002). Assessing Putative Human Pheromones. In: *Olfaction, Taste, and Cognition*, ed. C Rouby, B Schaal, D Dubois, R Gervais, & A Holley, pp. 178-195. Cambridge, U.K: Cambridge University Press.

Jaubert J N, Gordon G, & Doré J C (1987). Une Organization du Champ des Odeurs. *Parfum, Cosmétiques, Arômes* 77: 53-6.

Jaynes J (1976). *The Origin of Consciousness in the Breakdown of the Bicameral Mind.* Boston, MA: Mariner Books.

Kay P & McDaniel C K (1978). The Linguistic Significance and the Meanings of Basic Color Terms. *Language* 54: 610-646.

Kline N A & Rausch J L (1985). Olfactory Precipitants of Flashbacks in Post-traumatic Stress Disorder: Case Reports. *Journal of Clinical Psychiatry* 46: 383–4.

Knasko S C, Gilbert A N, & Sabini J (1990). Emotional State, Physical Well-Being, and Performance in the Presence of Feigned Ambient Odor. *Journal of Applied Social Psychology* 20: 1345–1357.

Köster E P (2002). The Specific Characteristics of the Sense of Smell. In: *Olfaction, Taste, and Cognition*, ed. C Rouby, B Schaal, D Dubois, R Gervais, & A Holley, pp. 27-44. Cambridge, U.K: Cambridge University Press.

Kraft P (2004). Aroma Chemicals IV: Musks. In: *Chemistry and Technology of Flavours and Fragrances*, ed. D J Rowe, pp. 143-165. Oxford: Blackwell.

Kubeczka K H & Formácek V (2002). *Essential Oils Analysis by Capillary Gas Chromatography and Carbon-13 NMR Spectroscopy*, 2nd ed. New York: Wiley.

Kurtz D B, White T L, & Hayes M (2000). The Labeled Dissimilarity Scale: A Metric of Perceptual Dissimilarity. *Perception and Psychophysics* 62: 152–161.

Larsson M & Bäckman L (1999). Implicit and Explicit Memory for Familiar and Unfamiliar Olfactory Information Across the Adult Life Span. *European Symposium on Olfaction and Cognition.* Lyon, France: French National Centre for Scientific Research (CNRS), June 10-12.

Lawless H T (1984). Flavor Description of White Wine by "Expert" and Nonexpert Wine Consumers. *Journal of Food Science* 49: 120–123.

Lawless H T (1989). Exploration of Fragrance Categories and Ambiguous Odors Using Multidimensional Scaling and Cluster Analysis. *Chemical Senses* 14(3): 349–360.

Le Guérer A (1992). *Scent: The Mysterious and Essential Powers of Smell.* New York: Random House.

Le Guérer A (2002). Olfaction and Cognition: A Philosophical and Psychoanalytic View. In: *Olfaction, Taste, and Cognition*, ed. C Rouby, B Schaal, D Dubois, R Gervais, & A Holley, pp. 3-15. Cambridge, U.K: Cambridge University Press.

Leitereg T J, Guadagni D G, Harris J, Mon T R, & Teranishi R (1971). Chemical and Sensory Data Supporting the Difference Between the Odors of the Enantiomeric Carvones. *Journal of Agriculture and Food Chemistry* 19(4): 785.

Linnaeus (1758). *Systema Naturae*. Sweden.

McCaffrey R J, Lorig T S, Pendrey D L, McCutcheon N B, & Garrett J C (1993). Odor-induced EEG Changes in PTSD Vietnam Veterans. *Journal of Traumatic Stress* 6: 213-24.

Miller G A (1956). The Magical Number Seven, Plus or Minus Two: Some Limits on our Capacity for Processing Information. *Psychological Review* 63: 81-97.

Mombaerts P (1999). Odorant Receptor Genes in Humans. *Current Opinion in Genetics and Development* 9: 315-20.

Moncrieff R W (1966). *Odour Preferences*. London: Leonard Hill.

Mookherjee B D & Wilson R A (1982). The Chemistry and Fragrance of Natural Musk Compounds. In: *Fragrance Chemistry: The Science of the Sense of Smell*, ed. E T Theimer. pp. 434-491. New York: Academic Press.

Morris E T (1984). *Fragrance: The Story of Perfume from Cleopatra to Chanel*. New York: Scribner.

Mouélé M, Hombert J M, Dubois D, Schaal B, Rouby C, & Secard G (1997). Specific Odor Terminology: The Case of Li Wanzi (a Bantu Language of Central Africa). *Chemical Senses* 20: 78.

Pandya V (1990). Movement and Space: Andamanese Cartography. *American Ethnologist* 17: 775–797.

Plato (1961). Timaeus. In: *The Collected Dialogues, Bollingen Series LXXI*, ed. E Hamilton & H Cairns, pp. 1149-1251. Princeton, NJ: Princeton University Press.

Plumacher M, Holz P, eds (2007). *Speaking of Colors and Odors: Converging Evidence in Language and Communication Research* 8. Amsterdam: John Benjamins.

Protectedstatic. (2012, November 11). These Diseases Can Be Diagnosed by Smell [Online forum comment]. Retrieved from http://io9.gizmodo.com/5959395/these-diseases-can-be-diagnosed-by-smell

Ravindran P N, Nirmal-Babu K, & Shylaja M, ed. (2003). *Cinnamon and Cassia: The Genus Cinnamomum, Medicinal and Aromatic Plants - Industrial Profiles*. Boca Raton, FL: CRC Press.

Ressler K, Sullivan S, & Buck L B (1994). Information Coding in the Olfactory System: Evidence for a Stereotyped and Highly Organized Epitope Map in the Olfactory Bulb. *Cell* 79: 1245-1255.

Rouby C & Bensafi M (2002). Is There a Hedonic Dimension to Odors? In: *Olfaction, Taste, and Cognition*, ed. C Rouby, B Schaal, D Dubois, R Gervais, & A Holley, pp. 140-159. Cambridge, U.K: Cambridge University Press.

Sachs C (1943). *The Rise of Music in the Ancient World, East and West*. New York: Norton.

Saha A, Panos Z, Hanna T, Huang K, Hernández-Rivera M, & Martí A A (2013). Three-Dimensional Solvent-Vapor Map Generated by Supramolecular Metal-Complex Entrapment. *Angewandte Chemie International Edition* 52(48): 12615–12618.

Sapolsky R (2010). The Limbic System. *Human Behavioral Biology*. Stanford: Stanford University course lecture, filmed April 30.

Schaal B, Rouby C, Marlier L, Soussignan R, & Tremblay R E (1998). Variabilité et Universaux de Espace Perçu des Odeurs: Approaches Inter-culturelles de L'hédonisme Olfactif. In: *Géographie des Odeurs*, ed. R Dulau & J R Pitte, pp. 24-47. Paris: L'Hartmattan.

Schab F R & Crowder R G (1995). *Memory for Odors*. Mahwah, NJ: Lawrence Ehrlbaum.

Schleidt M, Neumann P, & Morishita H (1988). Pleasure and Disgust: Memories and Associations of Pleasant and Unpleasant Odours in Germany and Japan. *Chemical Senses* 13: 279-293.

Sigma-Aldrich (2011). *Aldrich Chemistry 2012-2014: Handbook of Fine Chemicals*. Sigma-Aldrich.

Sigma-Aldrich (2013). GC Analysis of Sweet Orange Essential Oil. http://www.sigmaaldrich.com/. Accessed Dec 20, 2013.

Sperber D (1975). *Rethinking Symbolism*. Cambridge: Cambridge University Press.

Sperry R (1966). "Mind, Brain, and Humanist Values." Reprinted for private circulation from the *Bulletin of the Atomic Scientists*vol XXII no 7.

Stevenson R J, Prescott J, & Boakes R A (1999). Confusing Tastes and Smells: How Odours Can Influence the Perception of Sweet and Sour Tastes. *Chemical Senses* 24(6): 627-635.

Süskind P, trans Woods J E (1986). *Perfume: The Story of a Murderer*. New York: Vintage.

Turin L (2006). *The Secret of Scent: Adventures in Perfume and the Science of Smell*. New York: Harper Collins.

Wedekind C & Füri S (1997). Body Odour Preferences in Men and Women: Do They Aim for Specific MHC Combinations or Simply Heterozygosity? *Proceedings of the Royal Society B: Biological Sciences* 264(1387): 1471-9.

Weiss T, Snitz K, Yablonka A, Khan R M, Gafsou D, Schneidman E, & Sobel N (2012). Perceptual Convergence of Multi-Component Mixtures in Olfaction Implies an Olfactory White. *Proceedings of the National Academy of Sciences* 109(49): 19959-19964.

Wilson D A & Stevenson R J (2006). *Learning to Smell: Olfactory Perception from Neurobiology to Behavior*. Baltimore: Johns Hopkins University Press.

Zald D H & Pardo J V (1997). Emotion, Olfaction, and the Human Amygdala: Amygdala Activation during Aversive Olfactory Stimulation. *Proceedings of the National Academy of Sciences USA* 94: 4119-24.